P9-BZM-286

ISO 9000

Third Edition

ISO 9000

The Year 2000 and Beyond

Perry L. Johnson

Third Edition

McGraw-Hill

New York San Francisco Washington, D.C. Auckland Bogotá
Caracas Lisbon London Madrid Mexico City Milan
Montreal New Delhi San Juan Singapore
Sydney Tokyo Toronto

Cataloging-in-Publication Data is on file with the Library of Congress

McGraw-Hill

A Division of The ***McGraw-Hill*** *Companies*

2 3 4 5 6 7 8 9 DOC/DOC 0 9 8 7 6 5 4 3 2 1

ISBN 0-07-135655-X

The sponsoring editor of this book was Linda Ludewig. The editing supervisor was Steven Melvin, and the production supervisor was Pamela Pelton. This book was set in New Century Schoolbook per the MHT 6 x 9 design by Joanne Morbit of McGraw-Hill's Professional Book Group composition unit, Hightstown, N.J.

This book is printed on recycled, acid-free paper containing a minimum of 50% recycled, de-inked fiber

Printed and bound by R. R. Donnelley & Sons Company, Inc.

"You don't have to do this. Survival is not compulsory."

DR. W. EDWARDS DEMING
On the importance of ISO 9000

Contents

Part 2. Technical Requirements and Guidelines

Preface

The ISO 9000 quality system standard is much more than a major component of the worldwide drive for quality, for unlike other quality programs, practices, and procedures, ISO 9000 is international in scope and impact. Moreover, ISO 9000 has practical, dollars-and-cents implications for American business, in both the short and long run.

In the short run, implementing an ISO 9000 quality system has a major and positive impact on quality, productivity, and cost reduction. And ISO 9000 registration can give American businesses unmatched credibility and competitive advantages in the European Union.

In the long run, ISO 9000 implementation and registration will preserve and create domestic and international markets for American businesses in virtually every field. Even now, many major American businesses and government institutions are requiring ISO 9000 registration as a supplier quality assurance qualification.

The ISO 9000 standards first appeared in 1987 and were revised in 1994. For 2000, the ISO 9000 standards are undergoing a major overhaul.

This book was written to provide managers, engineers, and businesspeople of all disciplines, especially those who are not well-versed in quality technologies, with comprehensive information about the ISO 9000 standard, its applications, and the major changes it faces in the 2000 revision. Because the revision will not take effect until November or December 2000, this book offers transitional guidance.

The information in this book comes from a wide range of sources. The most important of these is the official *ISO 9000: International Standards for Quality Management*, published by the International Organization for Standardization (ISO) in Geneva, Switzerland, which also publishes drafts of the 2000 revisions, along with their final versions. Firms that are serious about ISO 9000 should rely on these documents. Please keep in mind, however, that all references in this book to the 2000 revisions are based on the latest available draft, which is subject to change. As a result, information in this book about the 2000 revisions is not definitive. The final version of the 2000 revisions is scheduled for release in November or December 2000.

Firms planning to adopt ISO 9000 also should obtain the aid of accredited ISO 9000 consultants in interpreting and implementing the standard to their specific situation. At Perry Johnson, Inc. (PJI), we have found a significant learning curve when a company attempts to implement ISO 9000 on its own. PJI has developed its patented A-to-Z implementation approach that fits ISO 9000 methodology to a client's processes.

Therefore, although much effort was put forth to make this book accurate and helpful, it should only be relied on as a general guide. Readers also should keep in mind that the 2000 revisions may experience major changes before they are finalized.

In addition, numerous articles and other printed material about the ISO 9000 standard were consulted. A list of these sources appears in the Bibliography.

Finally, the help of ISO 9000 consultants and implementers from North America and Europe is appreciated. Several dozen managers of American firms that have obtained ISO 9000 registration shared their experiences and views, and many of these appear in this book. For various reasons, some asked that their comments appear without attribution. The names of the rest appear in the Acknowledgments, and I am grateful to all.

Perry L. Johnson
Southfield, Michigan
December 1999

Acknowledgments

I would like to thank several people who helped make this book possible. Dave Hornstein played a major role in offering editorial contributions and reviewing the manuscript. He was ably assisted in gathering material for this book by Andrea Freeman.

The following individuals contributed to the creation of this book by participating in interviews, surveys, and other research efforts. Deepest thanks are extended for their cooperation.

Bradford Aho, Magnetic Data, Inc., Eden Prairie, Minnesota

Jeff Armstrong, H. P. Smith, Iowa City, Iowa

Mark Beale, Ceco Filters, Conshohocken, Pennsylvania

Dennis Beckley, Dayton-Rogers, Minneapolis, Minnesota

David Beckwith, Powdertech Corp., Valparaiso, Indiana

Craig Billings, Lectron Products, Bellefontaine, Ohio

Donald Blair, H. P. Smith, Bedford Park, Illinois

Keith Brown, Marmon/Keystone Corporation, Butler, Pennsylvania

Brian Burke, Container Products Corp., Southfield, Michigan

Bill Busher, Leybold Inficon, East Syracuse, New York

Dr. Wallace Carlson, University of Wisconsin-Stout

John Cline, Olympic Steel, Minneapolis, Minnesota

Anthony Coggeshall, Adhesives Research, Inc., Glen Rock, Pennsylvania

Mike Cook, Toppan Printronics, Dallas, Texas

David Coombes, United Kingdom

Art Coté, Delphian, Northvale, New Jersey

Frederick Douglas, Defense Research Inc., Anniston, Alabama

Joseph Druecker, Indiana Products Co., South Bend, Indiana

Jim Ecklein, Augustine Medical, Eden Prairie, Minnesota

Tim Edwards, CTS Corporation, Elkhart, Indiana

Steve Ekblad, Dayton Rogers, Minneapolis, Minnesota

Jesus Estrada, Keptel, Inc., El Paso, Texas

Charles Funk, Ingersoll-Rand Co., Davidson, North Carolina

Mark Gill, PPG Industries, Delaware, Ohio

Ray Grote, PMI Food Equipment Group, Troy, Ohio

Timmy Hale, First Chemical Corp., Pascagoula, Mississippi

Wayne Henderlong, Powdertech Corp., Valparaiso, Indiana

Tim Jestness, American Chrome & Chemicals, Corpus Christi, Texas

Linda Kabel, Menasha Corporation, Coloma, Michigan

Todd Krug, Skills Resource, Simpson, South Carolina

Doug Karns, Brush Wellman, Inc., Portage, Ohio

Sherman McDonald, GSE, Inc., Farmington Hills, Michigan

John Milsovic, H. P. Smith, Bedford Park, Illinois

Larry Montalto, Alusuisse, Bellwood, Illinois

Lorcan Mooney, United Kingdom

Charles Mortimer, Monarch Stamco, New Bremen, Ohio

Dennis Payne, Woodbridge Foam Corp., Woodbridge, Ontario, Canada

Bill Pietrzyk, USA-TACOM, Warren, Michigan

D.G. Prillaman, CTS Corporation, Elkhart, Indiana

Ray Querciagrossa, Allmand Industries, Inc., Livonia, Michigan

William Rahn, Ford Motor Company, Ypsilanti, Michigan

George Raub, TRW Vehicle Safety Systems Division, Washington, Michigan

John Renner, Sigma Stamping, Auburn Hills, Michigan

Mike Ruetz, Blackman Uhler Chemical, Spartanburg, South Carolina

Mel Sand, Federal Stampings, Rochester, New York

Julian Smith, General Motors Saginaw Division, Athens, Alabama

Tom Solomon, Federal Stampings, St. Petersburg, Florida

Mike Sweeney, CTS Corporation, Elkhart, Indiana

William Taylor, Keptel, Inc., Tinton Falls, New Jersey

John Troyer, Ford Meter Box Co., Wabash, Indiana
Dave Turteltaub, Phillips Circuit Assembly, Tampa, Florida
Don Van Hook, Strahman Valves, Inc., Florham Park, New Jersey
Mike Webb, Federal Stampings, St. Petersburg, Florida
Sandy Weller, Woodbridge Foam Corp., Woodbridge, Ontario, Canada
Noel White, United Kingdom
Wayne Williams, Kenmode Tool, Algonquin, Illinois
Stephen Wityl, Virginia Panel Corp., Waynesboro, Virginia
Larry Wojnicz, Outokumpu Copper, Kenosha, Wisconsin

ISO 9000 at a Glance

The ISO 9000 Series Standards

Type	Number	Title
Quality Management Models	ISO 9000-1: 1994	*Quality Management and Quality Assurance Standards—Part 1: Guidelines for selection and use*
	ISO 9000-2: 1993	*Quality Management and Quality Assurance Standards—Part 2: Generic guidelines for the application of ISO 9001, 9002, and 9003*
	ISO 9000-3: 1997	*Quality Management and Quality Assurance Standards—Part 3: Guidelines for the application of ISO 9001 to the development, supply, installation, and maintenance of software*
	ISO 9000-4: 1993	*Quality Management and Quality Assurance Standards—Part 4: Guide to dependability program management*
Contractual Models	ISO 9001: 1994	*Quality Systems—Model for quality assurance in design, development, production, installation, and servicing*
	ISO 9002: 1994	*Quality Systems—Model for quality assurance in production, installation, and servicing*
	ISO 9003: 1994	*Quality Systems—Model for quality assurance in final inspection and test*
Quality System Elements	ISO 9004-1: 1994	*Quality Management and Quality System Elements—Part 1: Guidelines*
	ISO 9004-2: 1991	*Quality Management and Quality System Elements—Part 2: Guidelines for services*
	ISO 9004-3: 1993	*Quality Management and Quality System Elements—Part 3: Guidelines for processed materials*

The ISO 9000 Series Standards (*Continued*)

Type	Number	Title
	ISO 9004-4: 1993	*Quality Management and Quality System Elements—Part 4: Guidelines for quality improvement*
Quality Management Guidelines	ISO 10005: 1995	*Quality Management—Guidelines for quality plans*
	ISO 10006: 1997	*Quality Management—Guidelines to quality in project management*
	ISO 10007: 1995	*Quality Management—Guidelines for configuration management*
Auditing Guidelines	ISO 10011-1: 1990	*Guidelines for Auditing Quality Systems—Part 1: Auditing*
	ISO 10011-2: 1991	*Guidelines for Auditing Quality Systems—Part 2: Qualification criteria for quality systems auditors*
	ISO 10011-3: 1991	*Guidelines for Auditing Quality Systems—Part 3: Management of audit programs*
Guidance Documents	ISO 10012-1: 1992	*Quality Assurance Requirements for Measuring Equipment—Part 1: Metrological confirmation system for measuring equipment*
	ISO 10012-2: 1997	*Quality Assurance Requirements for Measuring Equipment—Part 2: Guidelines for control of measurement processes*
	ISO 10013: 1995	*Guidelines for Developing Quality Manuals*
	ISO 8402: 1994	*Quality Management and Quality Assurance Vocabulary*

ISO 9000: 2000 Draft Standards

Number	Title
ISO 9000: 2000	*Quality Management Systems—Fundamentals and vocabulary*
ISO 9001: 2000	*Quality Management Systems—Requirements*
ISO 9004: 2000	*Quality Management Systems—Guidelines for performance improvement*
ISO 19011: 2001	*Guidelines on auditing quality and environmental management systems*

ISO 9000

Third Edition

Part

1

All About ISO 9000

Chapter 1

Today's Edge, Tomorrow's Imperative

If you do business in Europe, don't overlook new quality rules that take effect there on January 1 [1993]. . . . Businesses will have to meet [ISO 9000 international standards] in order to sell inside the European Community.

We're shooting for certification to ISO 9001. But we won't stop there. Once we've got 9001, we'll go after 9002. And then we'll knock off 9003 as well. We won't stop till we hit the top!

These quotes share three traits:

1. They concern ISO 9000, the international quality system standard.
2. Their sources are authoritative. The first appeared in a respected American business publication, and the second was stated by a prominent corporate CEO.
3. They are erroneous.

But this is how it has been with ISO 9000. Since 1989, when ISO 9000 first began to make itself felt in the American corporate consciousness, facts have been scanty and mingled with fiction. Assorted misnomers and misleading public accounts have created chaos for American businesses that operate, or intend to operate, in Europe.

At Perry Johnson, Inc. (PJI), an American-based ISO 9000 consulting and training organization, we responded to client appeals by seeking the expertise of our European associates, who have lived and worked with ISO 9000 and related quality system standards since the 1970s. In educating our clients, we aimed, in large measure, at dispelling myths about ISO 9000.

Here are just a few:

- *ISO 9000 is "a European standard."* This claim is sometimes made to make ISO 9000 seem alien, unintelligible, or inappropriate for American business. The facts are that ISO 9000 was derived from home-grown American quality standards that are still in use, the United States is a member of the International Organization for Standardization (ISO), and Americans sit on the committees responsible for creating and developing the series standards.
- *ISO 9000 is or will become "mandatory" in order to do business in Europe.* This is true for firms that make a relative handful of products, a list that may grow in the coming years. However, the doors to Europe did not slam shut on non-ISO-registered firms on January 1, 1993. ISO 9000 registration increasingly will become desired, expected, and perhaps even required in certain markets and industries, but this will be driven by customer requirements and competitive pressures, not by official or statutory mandate.
- *ISO 9000 is "a paperwork nightmare."* Ironically, this claim is made by American managers who think nothing of creating 500-page "quality manuals" and shelves of procedure books. Documentation is central to ISO 9000 requirements, for the purposes of planning, control, training, and providing objective/audit evidence of conformance. Its requirements are minimal, however, and in no way redundant. The typical ISO 9000 quality manual is 20 to 35 pages long.
- *ISO 9000 is "inspection-based."* This implies that ISO 9000 is a simple program to sort out defects. Not true. ISO 9000 requires that the quality system monitor conformance to requirements. This is just one small step in the continuous improvement cycle that is at the heart of the ISO 9000 quality system.
- *ISO 9000 is "weak on statistical techniques."* On the contrary, ISO 9000 does not require specific statistical techniques but does obligate management to implement statistical techniques where appropriate. This makes sense, because ISO 9000 is meant to be generic.
- *ISO 9000 "claims to guarantee quality out the door."* This is probably the most erroneous myth of all. No quality system or technique guarantees quality out the door. ISO 9000 is about meeting customer requirements with a system that is appropriate, planned, controlled, documented, and fully understood by all.

The mission of this book is to replace the myths with solid facts, such as these:

- Harmonized variants of the ISO 9000 standard have been adopted by more than 110 nations.
- Although registrations to ISO 9000 are, as a percentage of all eligible organizations, low in Europe and far lower in the United States, the trend is clearly on the rise. As of December 1999, there were 29,062 American ISO 9000 registrations. In December 1998, the last date for which world figures are available, there were 271,966 ISO 9000 registrations worldwide, including 24,987 in the United States. In January 1993, by contrast, there were only 27,616 worldwide registrations, with just 893 in the United States. Registration is growing rapidly because ISO 9000–registered firms are perceived to offer better quality and have a marketplace edge over unregistered competitors.
- Many U.S. government agencies, including the Defense Department, the Federal Aviation Administration (FAA), and the Food and Drug Administration (FDA), are in the process of harmonizing their quality requirements to the ISO 9000 standard or have already done so. In some cases, ISO 9000 has replaced agency quality requirements.
- Several industries have seized on the ISO 9000 quality concept by developing their own derivative standards. The automobile industry has developed such quality system standards as the Big 3 automakers' QS-9000 and TE Supplement, the German VDA 6.1, the French EAQF, and the Italian AVSQ, which have, in turn, led to the international ISO/TS 16949. The aerospace industry has produced AS9100, while the telecommunications field has developed TL 9000. More details about these derivative standards, particularly QS-9000, will appear throughout this book, primarily in Part 4.

These facts, and a broad base of international experience providing ISO 9000 consulting, training, and implementation services, lie behind the convictions that are at the heart of this book:

- By the end of the 1990s, international competitive and political pressures have caused ISO 9000 to become a major, if not preeminent, worldwide quality system, the international coin of quality for the year 2000 and beyond.
- The merits of the ISO 9000 quality system are such that even firms unaffected by competitive and/or political pressures would do well to adopt it.

Business managers across America need to understand ISO 9000, except, of course, those who do not care about survival.

What Is ISO 9000?

ISO 9000 is a series of quality assurance standards that were created by the International Organization for Standardization (ISO), based in Geneva, Switzerland. This organization is a consortium of virtually all the world's national standards bodies, from Albania to Zimbabwe, with 132 members. The U.S. representative to ISO is the American National Standards Institute (ANSI), a very familiar name in American industry.

The ISO was founded in 1946 to develop industrial standards that would facilitate international trade, a goal that has been realized in part through ISO 9000. The ISO 9000 series incorporates several guidelines, listed in "ISO 9000 at a Glance," which address everything from implementing sound auditing techniques to developing quality manuals. Organizations are registered to one of the three contractual models, ISO 9001, 9002, and 9003. Under the 2000 draft standards, organizations would only register to ISO 9001, which could be tailored to omit inapplicable requirements or limit the scope of application under Section 1.2, Permissible Exclusions.

ISO 9000 is not a product standard but a *quality system standard.* It applies not to products or services but to the process that creates them. It is designed and intended to apply to virtually any product or service made by any process anywhere in the world.

To achieve this generic state, ISO 9000 refrains, to the greatest extent possible, from mandating specific methods, practices, and techniques. It emphasizes principles, goals, and objectives. All these focus on one objective, the same objective that drives every business: *meeting customer expectations and requirements.*

A well-designed, well-implemented, and carefully managed ISO 9000 quality system provides confidence that the output of the process will meet customer expectations and requirements. It is aimed at providing that confidence to three audiences:

- The customers directly
- The customers indirectly, via third-party audits and quality system registration
- Company management and staff

ISO 9000 does so by requiring that every business activity affecting quality be conducted in a three-part never-ending cycle of *planning, control,* and *documentation.* This means that

- Activities affecting quality must be *planned* to ensure that goals, authority, and responsibility are defined and understood.
- Activities affecting quality must be *controlled* to ensure that specified requirements at all levels are met, problems are anticipated and averted, and corrective actions are planned and carried out.
- Activities affecting quality must be *documented* to ensure understanding of quality objectives and methods, smooth interaction within the organization, feedback for the planning cycle, and objective/audit evidence of quality system performance for those who require it, such as customers or third-party auditors.

ISO 9000 is not new or radical. It is good, hard-headed common business sense in codified, verifiable, and easily adapted form. It has much in common with other quality schemes, such as MIL-Q-9858A, W. Edwards Deming's 14 points, total quality management (TQM), and the Malcolm Baldrige National Quality Award criteria.

The main difference is that an organization can *register* to ISO 9000.

How ISO 9000 Is Applied

As detailed in Chap. 3, there are two broad ways to apply the ISO 9000 quality system standard. One way is to implement it for *quality management purposes,* i.e., to obtain its benefits for their own sake. The other way is obtain *registration,* also known as *certification,* to the ISO 9000 quality system standard. Registration provides the benefits of quality system management plus significant strategic advantages.

Some firms seek registration for competitive reasons or because they hope ISO 9000 registration will reduce the number of supplier quality assurance (SQA) audits they must undergo. Others register because theirs is a "regulated product" as defined by European Union product directives (see below). However, according to a study by Britain's authoritative Institute for Quality Assurance (IQA), most firms seek ISO 9000 registration because of pressure or requirements from customers.

Registration is awarded by an accredited third-party registration body, known as a *registrar,* after it satisfies itself, by reviewing documentation and conducting on-site audits, that the firm's quality system conforms to the ISO 9000 standard. ISO 9000 registration is renewable and enforced by surveillance audits by the registrar. Registration is surprisingly easy to get for firms that already have significant quality systems in place. Registration is also relatively easy to keep. But ISO 9000 is almost impossible to *fake.* It focuses on performance, documentation, and objective/audit evidence. There are no shortcuts. You must walk your talk.

As explained more fully in Chap. 3, the benefits of registration can be substantial. ISO 9000 registration affords access to markets, an enhanced quality image, and significant competitive advantages. It also has been shown to enhance business health. The IQA has found that the annual bankruptcy rate for nonregistered British firms is 7.14 percent, whereas the rate for registered firms is a mere 0.2 percent.

All these benefits notwithstanding, many American firms are after ISO 9000 registration due to international considerations. Let us take a look at those.

Impact of ISO 9000

ISO 9000 is an international standard with worldwide implications. However, since the most immediate and profound impact of ISO 9000 is on Europe, let us sketch out the political and economic stage there.

- The *European Union* (EU), previously known as the *European Community* and established in 1958, today consists of 15 member states: Austria, Belgium, Denmark, Finland, France, Germany, Greece, Ireland, Italy, Luxembourg, the Netherlands, Portugal, Spain, Sweden, and the United Kingdom. The EU comprises a marketplace of some 370 million people.
- The *European Free Trade Association* (EFTA) is a separate group of four nations: Iceland, Liechtenstein, Norway, and Switzerland.
- The *European Economic Area* (EEA) is the collective term for the EU and the EFTA. It comprises some 380 million consumers in 19 nations. Thus the EEA represents the largest single market in the world, larger than the United States and Japan combined.

Since adoption of the Single European Act in 1987, the EU, together with the EFTA, has been making strides toward unifying member nations into a single marketplace where goods, services, people, and capital can move freely, as within one country.

The single marketplace came into being in 1993. The EU is now positioned to establish common policy in a number of areas, including foreign trade, agriculture, competition, transport, research and technology, the environment, and a common currency, the euro.

Acting in concert, the EU Council of Ministers has adopted many policies and regulations governing the conduct of trade among member nations. Most particularly, the EU has begun requiring *product certification* for certain regulated products. Such products cannot be marketed in the EU without independent certification as to safety and other factors. Approved products bear CE marking. This is a type of passport that allows a company's goods to cross into European countries.

The EU also has established guidelines for *quality systems registration.* To be certified, a regulated product must be made under the aegis of a quality system that has been audited and certified independently as meeting EU-approved quality system standards. These standards, which the EU calls *EN 29000,* are in fact an exact harmonized equivalent of ISO 9000.

In short, ISO 9000 is the recognized, accepted, and for certain products, mandated quality system standard in Europe. Although the doors did not slam shut on unregistered firms on January 1, 1993, the time is rapidly coming when ISO 9000 will be the rules by which the quality game is played in the world's largest marketplace.

Chapter

2

From America and Back: The Origins of the ISO 9000 Standard

ISO 9000 is more than a quality system; it is a total management system. It is the only total quality system that works.

ANTHONY COGGESHALL
Adhesives Research, Inc.
Registered to ISO 9001

The ISO 9000 quality system standard is still somewhat new to American businesspeople. However, most U.S. quality professionals understand the concept of quality system standards. First- and second-tier suppliers of major manufacturers such as the Big 3 automakers, as well as primary Defense Department contractors, have been subject to quality systems and audits for many years.

Some of these programs are similar to ISO 9000. As Jim Ecklein of Augustine Medical, a medical devices manufacturer registered to ISO 9001, notes, "The FDA has long mandated a Good Manufacturing Practices program, which is very compatible with ISO." Augustine Medical also has become registered to EN 46001, the European standard for medical device suppliers, and the EU Medical Device Directive.

George Raub, whose TRW Vehicle Safety Systems Division was registered in 1993, has since had two other divisions registered, one to ISO 9001 and the other to QS-9000. He notes that ISO 9000 has many similarities to the General Motors Targets for Excellence and the Ford Motor Company Q1 programs.

These resemblances are no accident. ISO 9000 is an international standard, sponsored by a Swiss-based organization and the direct descendent of a British quality system. Like many other quality-oriented techniques such as statistical process control (SPC), however, the ISO 9000 standard has directly traceable American ancestry. Moreover, the United States, through the American National Standards Institute (ANSI), is a member nation of ISO, and Americans participated in formulating the ISO 9000 standard. Thus ISO 9000 can by no means be considered foreign by American businesspeople.

ISO itself states that its series of international standards "embodies a rationalization of the many and various national approaches" to quality systems. Even so, there are very real differences between the application of quality systems in the United States and in Europe. These differences are crucial to understanding the role of ISO 9000 today. In this chapter we will trace the history and development of

- Quality standards generally
- ISO 9000's distant forebears
- ISO 9000's parents
- ISO 9000's siblings and sponsors

Standards: From Product to Quality to System

No one knows who invented the first set of standards. Their development, however, is a key to the emergence of the high-technology age as we know it. Technological advancements and the development of standards have gone hand in hand.

Centuries ago, craftspeople made products on a highly individualized basis. Operating in cottage settings, they employed unique and highly personalized design standards. Their standards for quality were likewise individualized. Typically, quality standards were driven by individual, one-on-one customer feedback, a method that remains highly pertinent today.

In medieval times, craftspeople began to band together to form guilds and trade unions. These groups created their own standards by which expertise in the various skills was measured. At the same time, military organizations found that the quality of the equipment and materials they used was quite literally a matter of life and death. As a result, quality standards were imposed on suppliers.

A thousand years ago, the English king appointed an officer to oversee the production of naval ships. Another official was put in charge of the quality and effectiveness of land-based weaponry and engineering.

Standardization itself became preeminent with the dawn of the industrial age. Eli Whitney may be most famous for his cotton gin, but the impact of this achievement pales compared with his use of standard, interchangeable parts in the making of rifles. Military applications such as these continued to drive the development and refinement of product standards and eventually quality standards as well.

By the start of the twentieth century, quality standards were being institutionalized and documented. As far back as 1912, the British government created an office to ensure the quality of military aircraft.

American Quality Standards

Some pinpoint the dawn of America's quality age to the recession of 1981–1982, the emergence of Japan as an economic threat, and the appearance of "Quality Is Job 1" as an advertising slogan of the Ford Motor Company.

To the contrary, documented quality standards of various kinds have been a fact of life in the American military establishment and among defense contractors since World War II, when the federal government established the MIL STD series of standards. These standards were, as we shall see, a key to the development of international quality standards as we know them today. Arguably, the most significant of these standards is MIL-Q-9858, a quality management program adopted in 1959 by the U.S. Department of Defense. Succeeded by MIL-Q-9858A in 1963, this standard was imposed on most of the department's suppliers for decades. A significant change has occurred, however.

In 1994, this long-time standard was eliminated for new department contracts, and then it was phased out in 1996. Instead, the Defense Department, along with the National Aeronautics and Space Administration (NASA) and the General Services Administration (GSA), three of the largest government purchasers, began promoting the use of ISO 9000 as a government-wide quality standard for purchasing.

All three of these federal agencies are committed to the use of ISO 9000 in varying degrees. This government agency movement to use ISO 9000 standards instead of government quality requirements stems from the cost savings of using a single quality standard for suppliers, instead of using multiple quality systems, and the growing worldwide acceptance of ISO 9000.

A parallel development that has shaped the American quality age was the creation of supplier quality standards by such major manufacturers as the Big 3 automakers and first-tier Defense Department suppliers.

These organizations, which rely heavily on suppliers for subassemblies and components, realized that meeting their own quality needs required them to impose quality standards on their suppliers.

This resulted in proprietary quality standards, such as the Ford Motor Company Q1 and the General Motors Targets for Excellence. Gradually, the requirements of such standards steadily tightened as they became more demanding and specific as to the types of quality tools, techniques, and reporting systems suppliers had to use.

This process led to the creation of the QS-9000 standard, an automotive sector–specific harmonized version of ISO 9000 for Big 3 suppliers, and its Tooling and Equipment (TE) Supplement, a harmonized version of ISO 9000 for automotive tooling and equipment suppliers. In Europe, ISO 9000 was the basis for three automotive quality standards, VDA 6.1 in Germany, EAQF in France, and AVSQ in Italy. Conflicts among these standards caused difficulties for international automobile suppliers and has led, in turn, to the development of ISO/TS 16949, an international automotive standard.

Quality Standards Overseas

Overseas, particularly in the United Kingdom, the development of quality systems standards took a somewhat different course.

AQAP, DEF/STAN, BS 5750...

If imitation is the sincerest form of flattery, then MIL-Q-9858A is a flattered program indeed. In 1968, the North Atlantic Treaty Organization (NATO) adopted it as its own quality system standard under the name AQAP-1 (Allied Quality Assurance Publication 1). The British Department of Defense, knowing a good thing when it saw one, incorporated most of AQAP-1 provisions 2 years later in its own quality system standard, DEF/STAN 05-8.

It is generally assumed that U.S. quality methods and standards are 5 to 10 years ahead of the British and that the United States has much more of a laissez-faire tradition than Great Britain. However, the British were the first to go beyond implementing official quality standards for defense alone. This early and intense interest in quality as a policy imperative may have stemmed from the distance British quality had to travel.

In any event, in 1979 Great Britain took the unprecedented step of officially adopting a quality system standard for its commercial establishment. In accordance with the philosophy of Prime Minister Margaret Thatcher, "Quality is about making products that don't come back for customers that do," the British Standards Institution, at the time a government agency, developed BS 5750, a quality system standard evolved from DEF/STAN, AQAP, and MIL-Q-9858A.

The British government not only created the standard but also actively supported and promoted it. The Department of Trade and Industry, equivalent to the U.S. Department of Commerce, sponsored the United Kingdom Accreditation Service (UKAS), which accredits quality system registrars, and the International Register of Certificated Auditors (IRCA), which accredits quality system auditors and auditor training. These matters are discussed in more detail in Chap. 13.

Even more important, the British government vigorously promoted BS 5750 throughout the private sector. It actively encouraged firms to register and publicized BS 5750 to increase consumer awareness and acceptance of the standard.

...EU, ISO, and EN

In 1987, probably the most important year in the history of postwar Europe, the Single European Act was passed by the European Union. This set a course intended to lead to the development of an economically, if not politically, united Europe.

That same year saw these landmark advances in the development of quality system standards:

- The International Organization for Standardization (ISO) created the ISO 9000 quality system standard, drawing on the British BS 5750 standard.
- The British harmonized BS 5750 with ISO 9000.
- The British modified DEF/STAN 05-8 to harmonize it with ISO 9000.
- The European Union (EU) adopted the EN 29000 quality standard, which is virtually identical to ISO 9000.

The differences between American and European applications of quality system standards are clear. In the United States, quality standards mainly have been driven by individual corporations and military imperatives. In Europe, governments and the EU, which is becoming a supergovernment, have been much more actively involved. The role of government has critical implications for U.S. firms wishing to register to ISO 9000, as discussed in Chap. 13.

Most industrialized nations have adopted harmonized versions of ISO 9000. No doubt for reasons of nationalist pride, these nations have given their standards their own designations, even though the standards themselves are ISO 9000, word for word. For example, ISO 9000

TABLE 2-1. Harmonized Versions of ISO 9000

Nation	Standard
Australia	AS 3900
Austria	ÖNORM EN 29000
Belgium	NBN-EN29000
Canada	CAN/CSA-ISO 9000
China	GB/T 19000
Denmark	DS/ISO 9000
Finland	SFS-ISO 9000
France	NF-EN 29000
Germany	DIN ISO 9000
Hungary	MSZ EN 29000
India	IS 14000
Ireland	IS/ISO 9000/EN 29000
Italy	UNI/EN 29000
Japan	JIS Z 9900
Malaysia	MS-ISO 9000
Netherlands	NEN ISO 9000
New Zealand	NZS 9000
Norway	NS-ISO 9000
South Africa	SABS/ISO 9000
Spain	UNE 66 900
Sweden	SS-ISO 9000
Switzerland	SN EN 29000
Tunisia	NT 110.18
United Kingdom	BS 5750
United States	ANSI/ASQ Q9000

is still referred to as BS 5750 in the United Kingdom, NS-ISO 9000 in Norway, and CAN/CSA-ISO 9000 in Canada (see Table 2.1).

ANSI/ASQ Q9000 is the American version of ISO 9000. However, here, the standard's sponsorship and application are quite different.

ISO 9000 in America

As noted earlier, quality in the United States traditionally has been a private-sector process driven by the customer-supplier relationship. Compared with European governments, the U.S. government has had only sparse involvement in quality. It sponsors the Malcolm Baldrige National Quality Award and promotes the total quality management (TQM) concept and other quality initiatives through the National Institute of Standards and Technology (NIST). But the U.S. government does not mandate quality systems, sponsor training and accreditation organizations, or use its "bully pulpit" to promote quality awareness to the public.

ISO 9000's American variant developed without American government sponsorship. The American Society for Quality (ASQ), a professional society, jointly with the American National Standards Institute (ANSI), which represents the United States on ISO, publishes a harmonized variant of the ISO 9000 standard under the name Q9000, which is not to be confused with QS-9000.

ASQ, ANSI, and Q9000 have, at this writing, not received official U.S. government sponsorship.

ISO 9000: Common Ground or Battleground?

Are American quality standards stricter than European standards? Is U.S. quality better? In many respects, the answer to both questions is yes.

However, ISO 9000 is by no means a foreign standard, for its American parentage is clear. The real issue is leadership. The EU and ISO have done something the United States has not. By adopting ISO 9000 and EN 29000, raising their visibility, publicizing them, and mandating their use in critical industries, the EU has made quality a matter of national and international policy.

Moreover, the EU has created a comprehensive yet relatively simple, straightforward, and most of all, generic quality system standard. U.S. quality systems are, on the other hand, by no means universal. They affect some industries heavily, with sometimes almost ruinous redundancy, and leave others untouched.

American quality standards tend to stress manufacturing and quite often mandate specific tools and techniques. And with rare exceptions, such as Ford's "Quality Is Job 1" campaign, U.S. quality programs and standards are almost unknown to the general public.

As we will see, ISO 9000 is general enough to apply to virtually any product, service, or facility yet specific enough to create genuine and continuous improvement. It is oriented toward results rather than technique. It affects every process, employee, and function in the firm.

In Europe in general and in Great Britain in particular, ISO 9000 and its equivalents enjoy strikingly high public awareness and acceptance. These are the attributes that give the ISO standard its power and make it a force with which U.S. firms must reckon as they strive to maintain and build business in the EU.

Some have deemed ISO 9000 an attempt by the EU to erect trade barriers. Ultimately, however, it is futile and self-defeating to treat ISO 9000 as a battleground. ISO 9000 is in fact an *opportunity* for American businesses—an opportunity to obtain benefits, most practical, some measurable.

We turn to these benefits in the next chapter.

Chapter

3

Benefits of ISO 9000

What is this, another "program of the month"?

Oh, goody. More audits.

What can these Europeans teach us about quality that we don't already know?

Americans react to ISO 9000 in various ways. In large part, they are skeptical. The main concern is a very logical one: "What's in it for me?"

For many firms, the answer is obvious. They know for a fact that without ISO 9000 they will lose business. If your firm is in this position, this threat may manifest itself in various ways:

- The European Union (EU) makes ISO 9000 registration mandatory for firms in your marketplace.
- Your biggest customer requires ISO 9000 of all suppliers.
- Your major competitor has adopted ISO 9000.

Protecting existing markets is just one of the many practical and measurable benefits that has drawn American firms in droves into the realm of ISO 9000. Other benefits are less measurable, more philosophical, and definitely oriented to the long term. They lack the dramatic immediacy of hammering the competition or saving the firm from oblivion. However, they are just as profound as the more immediate benefits.

The ISO 9000 quality system standard has both short- and long-term implications, with both tactical and strategic applications. As the standard itself says, in ISO 9000-1: 1994, *Quality Management and Quality Assurance Standards—Part 1: Guidelines for selection and use,* Section 6, its objectives affect competitiveness, as well as quality:

> The supplier's organization should install and maintain a quality system designed to cover all the situations that the organization meets....this system will strengthen its own competitiveness to fulfill the requirements for product quality in a cost-effective way.

In this chapter we answer the question, "What's in it for me?" by examining the benefits of ISO 9000. To a great extent, the range and intensity of the benefits are determined by the way the standard is applied. The structure of ISO 9000 itself acknowledges this. It provides for two broad avenues of application:

- *Quality management purposes,* in which the facility adopts the standard as a blueprint for its internal quality system
- *Contractual purposes,* in which a demonstrated quality system is a condition of a contract with a customer

The benefits obtained through contractual application of ISO 9000 include the practical and immediate advantages cited above, as well as the general and long-term benefits obtained through developing and operating a genuine quality management system. Let us look at the quality management benefits first.

What Is a Quality System?

The aim of a quality system is to ensure that the facility's product or service, generically referred to as *output,* meets the customer's quality requirements. The quality system incorporates both quality assurance and quality control.

Fine. But what is *quality?* What, for that matter, are *quality assurance* and *quality control?*

The ISO 9000 standard is carefully designed. Each element is meticulously defined. Each section of the standard offers thorough definitions of the terms used, and the definitions are also gathered and presented in ISO 8402: 1994, *Quality Management and Quality Assurance Vocabulary,* which will be replaced by ISO 9000: 2000, *Quality Management Systems—Fundamentals and vocabulary.* Let us look at how the standard defines these critical terms.

Quality, as ISO 9000 interprets it, is an integration of the features and characteristics that determine the extent to which output satisfies the *customer's* needs. Notice that this definition is very customer-driven. The customer determines what features and characteristics are important. Customers judge the extent to which the features and characteristics of the output satisfy their needs. ISO 9000: 2000, Section 2.1.1 makes only slight changes by defining *quality* as the "ability of a set of inherent characteristics of a product, system or process to fulfill requirements of customers and other interested parties."

Quality assurance is the collective term for planned and formalized activities intended to provide confidence that the output will meet required quality levels. In addition to in-process activities, quality assurance includes an array of activities external to the process, such as those undertaken to determine customer needs. ISO 9000: 2000, Section 2.2.11 defines *quality assurance* as the "part of quality management focused on providing confidence that quality requirements are fulfilled."

Quality control is the collective term for in-process activities and techniques intended to create specific quality characteristics. These activities include monitoring, reduction of variation, elimination of known causes, and efforts to increase economic effectiveness. ISO 9000: 2000, Section 2.2.10 defines *quality control* as the "part of quality management focused on fulfilling quality requirements."

A quality system, then, is a management-driven, facility-wide, and process-wide program of plans, activities, resources, and events. This program is implemented and managed with the aim of ensuring that process output will meet customer quality requirements, logically ensuring that return-on-investment (ROI) goals are met.

An effective quality system is the philosophical and procedural glue that unites all elements of the facility, including employees, plant, equipment, and procedures, with *suppliers* at the input end and *customers* at the output end.

Facilities that operate quality systems tend to exhibit the following attributes:

- A philosophy of prevention rather than detection
- Continuous review of critical process points, corrective actions, and outcomes
- Consistent communication within the process and among facility, suppliers, and customers
- Thorough recordkeeping and efficient control of critical documents
- Total quality awareness by all employees
- A high level of executive management confidence and support

These attributes inevitably lead to the following tangible benefits:

- Informed and competent management decision making
- Dependable process input (supplier control)

- Control of quality costs
- Increased productivity
- Reduced waste

In sum, the facility with a well-designed and properly implemented quality system has a process that tends to be lean, sensitive to customer needs, highly reactive, efficient, and positioned at the leading edge of its marketplace.

These are the benefits obtainable by facilities that implement ISO 9000 for quality management purposes. Implementation is achieved by following the guidelines in ISO 9000-1: 1994 or ISO 9000-2: 1993, which will be replaced by ISO 9000: 2000. These components of the ISO 9000 standard are defined in Chap. 4.

Most often, facilities get involved with the ISO 9000 quality system standard for *contractual* purposes, because a customer has specified an ISO 9000 quality system as a condition of the contract. Such a condition usually requires the facility to become *registered* to the standard.

ISO 9000 Registration

Facilities that register to ISO 9000 obtain all the benefits of quality management users, plus a few more. Before detailing these, let us take a brief look at the ISO 9000 registration process, which is examined in detail in Chap. 14.

A facility seeks *registration* to ISO 9000 for one or more of the following reasons:

1. One or more customers require it by contract.
2. The facility expects such contractual requirements to be imposed at some point.
3. The facility views the registration approach as the most logical and effective way to implement and manage the quality system.

Unlike quality management applications of ISO 9000, which do not involve firms outside the facility, registration requires involvement with *registrars,* who are outside agents. Registrars are specially accredited for this purpose, and the process of selecting the correct one for your needs can be tricky. The selection process is detailed in Chap. 14.

Multiple facility registrations, also known as *blanket, umbrella,* or *corporate certifications,* are permitted under European Cooperation for Accreditation (EA) guidelines under specific circumstances. A company operating through a number of outstations or separate facilities may register the entire entity as a whole on one certificate, provided that all outstations

- Are part of the same entity
- Are under the same control
- Are doing substantially the same job
- Are under common management
- Use the same quality system and procedures

The advantage to multiple-facility registration is the lower costs involved. The corresponding disadvantage is that if one facility experiences implementation difficulties, the entire corporation's registration certificate is endangered. Where multiple facilities exist but separate certificates are sought, some of the efficiencies of the multiple-facility registration scheme can be realized without endangering company-wide registration.

Registration is conferred when the registrar satisfies itself, by means of process and documentation audits, that the facility

- Has a quality system that meets the ISO 9000 standard
- Uses that system actively in its daily course of activities

The facility becomes registered to one of the three ISO 9000 standard quality system models, ISO 9001, 9002, and 9003, using the model that most closely fits the scope of its operations. Under the 2000 drafts, facilities will only be registered to ISO 9001, which could be tailored to fit their scope of operations. These matters will be described in detail in Chap. 4.

The facility's quality system need only include those elements of the standard which are relevant to its effective operation. The system must be documented by means of one or more levels of documentation, including, for most facilities, a quality manual as the top tier.

Registration, once awarded, is reinforced and enforced by means of on-site audits. These surveillance audits review any changes to the quality system and ensure that corrective actions required by previous audits have been carried out.

Benefits of registration

One benefit of registration is that the facility regularly undergoes objective audits by skilled outside quality professionals. This alone is a powerful argument for registration. There are three other compelling reasons as well: access to markets, competitive issues, and potential net reduction in audits.

Access to markets. Access to markets is the most critical benefit of ISO 9000 registration. It enables facilities to maintain or create customer relationships in situations where ISO 9000 registration is required.

As noted in Chap. 1, the EU Council of Ministers now mandates ISO 9000 registration for makers of certain types of products. These include commercial scales, construction products, gas appliances, industrial safety equipment, medical devices, and telecommunications terminal equipment.

More products may be added to this list via additional product directives. This is especially likely for products and services that are potentially hazardous or involve personal safety or which are otherwise affected by product liability or similar regulations. Similarly, General Motors and DaimlerChrysler require QS-9000 registration for their tier-one suppliers, with DaimlerChrysler requiring TE Supplement registration for its tooling and equipment suppliers.

More major firms in Europe and elsewhere are moving toward mandating ISO 9000 registration by their suppliers. Dennis Beckley of Dayton-Rogers, whose facility has been registered to ISO 9002, found this to be the case. "One customer said to us, 'I'm being forced to find ISO 9000–registered vendors. Once I find one, it's going to get all my business.' " Woodbridge Foam Corp. experienced similar pressure on its way to ISO 9002 registration. "One of our biggest customers," says Sandy Weller, "asked about ISO awhile back. Now they're requiring it. They gave us a year to get registered." And Anthony Coggeshall of Adhesives Research, which has a major commercial presence in the EU and has registered to ISO 9001, saw ISO pressure coming long ago: "Our customers started asking for it as far back as late 1990 and early 1991." For firms like these, the benefit of registration is that it enables them to retain their existing markets.

Competitive issues. Other firms have sought ISO 9000 registration because of *competitive threats.* Take Menasha Corporation. "One of our competitors," says Linda Kabel, "let it be known that it will be certified by the end of next year. So now our customer wants to know if we are going to be certified too."

Ironically, Kabel was already in the process of implementing an ISO 9001 quality system. "We were using it as a guideline to setting up a whole quality process," says Kabel. "But now there's all kinds of urgency. Our sales manager is putting on major pressure, saying we could lose our customer if we don't get certified."

There is commercial pressure arising simply from stiffening international competition and the needs of firms everywhere to differentiate among their suppliers. George Raub of TRW asserts: "One of our major overseas accounts asked very pointed questions about our plans for ISO 9000." Don Van Hook of Strahman Valves, which achieved ISO 9001 registration in 1995, describes the inquiries of one of the firm's European distributors: "They're selling in the EU, and they're thinking they'll have to require certification sometime. We know they're very sensitive to the issue of ISO certification. So we moved on it." Monarch Stamco registered to ISO 9001 in 1993 simply because the ISO 9000 standard was receiving greater and greater attention. "I'm seeing ISO 9000 mentioned more and more on RFQs," says Charles Mortimer.

Perhaps your facility does not fall into these categories. If this is the case, then the major benefit of registration may be the clear *competitive edge* it will give you over facilities that are not registered.

Registered facilities are authorized to display a special mark or logo. EU firms understand and value the significance of this mark. As quality becomes an increasingly vital marketplace distinction, ISO 9000–registered facilities will enjoy a clear competitive advantage.

Potential audit reduction. The final benefit of ISO 9000 registration, and for the most part merely a speculative one at this point, is the potential for reduced audits. Many facilities in certain industry segments undergo dozens of customer quality audits each year, some as many as 30 per month.

As ISO 9000 registration becomes understood and accepted in the United States, it is possible and perhaps even likely that many customers will accept current ISO 9000 registration in lieu of site audits, mail-in audits, or other redundant supplier quality assurance programs. This is certainly the hope of Brian Burke of Container Products. "We believe that once our quality system is ISO certified, we'll be exempted from some of our larger customers' supplier audits."

ISO 9000 registration, then, gives the facility the benefit of an objectively evaluated and enforced quality system. It offers the potential of reducing time-consuming and expensive supplier audits in the future. And it is a powerful strategic benefit for facilities having current or planned business ties with the EU, including

- Facilities located in the EU, actively making and/or marketing products or services there
- Facilities with corporate ties to firms making and/or marketing products or services in the EU
- Facilities that export to the EU

The bottom line

ISO 9000 is an ideal quality system for facilities that are serious about quality. It is an emerging imperative for any facility that has, or expects to have, commerce with EU nations. And registration provides the EU-hungry facility with an incomparable competitive edge.

These, along with the quality system benefits cited earlier, make achievement of ISO 9000 standards a powerful strategic tool, whether the facility goes the registration route or not.

Many firms, such as Augustine Medical, have already benefited from registration. As Jim Ecklein notes: "It's opened a lot of doors for us. Our European theater has taken off like gangbusters." Anthony Coggeshall of Adhesives Research concurs: "It's opened doors to sales that were closed to us before, especially some of the medical markets to whom we were simply an unknown. That changed once they saw that we were ISO registered."

Voices of the Users: What ISO 9000 Achieves

> We expect to see significant productivity improvements.
>
> Art Coté, Delphian, Northvale, NJ

> [We expect ISO to give us] a competitive edge.
>
> Wayne Williams, Kenmode Tool, Algonquin, IL

> We're looking for better control and to retain customers who require ISO 9000.
>
> Mike Sweeney, WTS Corporation, Elkhart, IN

> [ISO 9000] improves productivity, reduces costs, reduces process scrap. We expect to earn additional world market share.
>
> John Troyer, Ford Meter Box Co., Wabash, IN

> We hope ISO 9000 will help us get easier certification by customers.
>
> William H. Taylor III, Keptel Inc., Tinton Falls, NJ

> We're seeking ISO 9000 registration in order to continue business in Europe.
>
> Stephen F. Wityl, Virginia Panel Corp., Waynesboro, VA

> We are pursuing certification because we expect a higher level of access and credibility with our customers, current and potential, who market in Europe.
>
> Mike Ruetz, Blackman Uhler Chemical, Spartanburg, SC

> ISO 9002 has helped us improve our advance quality planning by providing us with a more thorough means of planning our processes and defining what resources will be needed.
>
> John Renner, Sigma Stamping, Auburn Hills, MI

> ISO 9002 has helped us to tighten up our process by demanding accountability and allowing us to provide directions to our employees.
>
> Todd Krug, Skills Resources, Simpson, SC

The testimony of satisfied users continues. And with it, the list of potential benefits of ISO 9000 registration grows.

Chapter

4

Components of the ISO 9000 Standard

Let's face it: Nobody likes quality. Quality in America bears the stigma of being the industrial cop. For all the preaching about quality, it is still seen as a necessary evil by American manufacturing. But ISO 9000 is different.

ANTHONY COGGESHALL
Adhesives Research, Inc.
Registered to ISO 9001

ISO 9000 to me is 80 percent psychology and 20 percent technique. Rather than force it down the throat of a reticent individual, the resistance goes when you think of it as "How can we bring ISO 9000 into our system?" In that awareness, the fear disappears.

DR. WALLACE CARLSON
University of Wisconsin–Stout

ISO 9000 has become a sweeping generic term, a topic, and a subject heading. As the offspring of many parents, it has been reviewed, revised, amended, and corrected by many over the years, with the 2000 drafts offering a complete restructuring. This constant overhauling has made ISO 9000 almost, but not quite, a "horse designed by committee." It also has made the standard and its components less than user-friendly and led to a certain level of confusion among those seeking information on the subject. The more streamlined 2000 drafts seek to be more user-friendly and less confusing.

Moreover, since the current ISO 9000 boom began in 1991, the topic has been written about and discussed by many people and publications all over the world. Unfortunately, the volume of editing, revision, discussion, and in some cases, poorly informed opinion has led to a certain amount of imprecision and error with respect to the standard. To illustrate the point, two more myths can be dispelled:

- ISO 9000 does have a precise meaning beyond its use as a generic term or subject heading. It is the official designation of the overseeing document that describes the quality system standard (see below).
- ISO 9000 registration is a bit of a misnomer. A facility does not become registered to ISO 9000. It becomes registered to one of the three contractual quality system models: ISO 9001, 9002, or 9003. Under the 2000 revisions, facilities will only be registered to ISO 9001.

To remove confusion, let us go straight to the source: the array of documents embodying the ISO 9000 standard, a list that is greatly reduced under the 2000 revisions. The parent documents are published and distributed by ISO. Harmonized variants also are published under the various national schemes listed in Chap. 3. These documents describe the elements of the quality system standard in detail, including

- Quality system requirements
- Quality system recommendations
- Guidelines for selection and use

The documents also provide supporting information, such as definitions of terms, guidelines for auditing, standards for measuring equipment, and rules applying to registrar accreditation. As part of the 2000 revision process, these documents are being reviewed by ISO Technical Committee (TC) 176 for incorporation within the revised standards, withdrawal, or reissue as technical reports.

This chapter examines the structure of the ISO 9000 quality system standard as it is presented by the documents that embody it.

Quality System Models

Three documents, ISO 9001, 9002, and 9003, are the heart of the quality system standard as it is commonly applied today. When the 2000 revisions take effect, only ISO 9001 will remain. The three documents apply to facilities that seek ISO 9000 registration for contractual purposes, where customers have made adherence to an ISO 9000 quality system a condition of doing business.

Sometimes the contract specifies the model to which the facility must adhere. In such cases, the facility has no trouble deciding to which model it must conform. At other times, no specific contractual requirements are in place, but the facility anticipates a contractual obligation. In such a case, the facility first selects the model that most closely fits the scope of its operation. Under the 2000 drafts, ISO 9001 can be tailored to fit a facility's scope of operations under Section 1.2, Permissible Exclusions.

The facility then

- Implements a quality system that conforms to the requirements of the model
- Obtains registration, independent and verifiable evidence that the facility's quality system is consistent with and meets the requirements of that model

The contents of the contractual models are described in detail near the end of this chapter. Implementation of ISO 9000 quality systems and registration to the standard are covered later in this book.

Quality Management Guidelines

ISO 9000 is not just for the facility that is or expects to be under contractual obligation. The standard provides guidance for the facility wishing to implement an ISO 9000 quality system for its inherent benefits. Such facilities use ISO 9000's *quality management guidelines.*

These documents, ISO 9004-1, 9004-2, 9004-3, and 9004-4, which are to be replaced by ISO 9004, *Quality Management Systems—Guidelines for performance improvement,* under the 2000 drafts, present quality system models. But facilities do not register to these guidelines. Quality systems based on them are not subject to audit. However, the guidelines simply spell out comprehensive quality management systems adaptable by virtually any organization.

The quality management and quality system guidelines serve another important purpose. ISO 9004-1, 9004-2, 9004-3, and 9004-4 can and should be consulted by facilities developing quality systems based on the contractual models. Once the 2000 revisions take effect, ISO 9004 should be consulted. Many of the passages in the guidelines serve to clarify and expand on the relatively pithy provisions of the contractual models. The guidelines themselves are not inherently enforceable as a condition of registration to any of the contractual models.

Guidelines for Auditing

An essential element of an ISO 9000 quality system is the continuous gathering and evaluation of objective/audit evidence about the performance of the system against specified requirements. One way this evidence is obtained is through audits. There are three kinds of audits:

- *First-party,* or *internal, audits,* carried out by facility personnel in accordance with the facility's quality policy. Internal audits are required by the contractual models.
- *Second-party audits,* wherein a customer audits the facility's quality system. These audits were a common part of the American quality scene even before the advent of ISO 9000.
- *Third-party audits,* which are usually carried out by an accredited third party, often a registrar, to provide objective/audit evidence, followed by a seal of approval, that the facility's quality system meets the published standards. In the context of this discussion, audit approval results in *registration* to the ISO 9000 quality system standard.

ISO publishes a series of three documents, the 10011 series, *Guidelines for Auditing Quality Systems,* which governs the conduct of audits. These will be replaced by ISO 19011, *Guidelines on Auditing Quality and Environmental Management Systems,* under the 2000 revisions.

- ISO 10011-1: 1990, *Auditing,* establishes basic audit principles, criteria, and practices. It spells out a system governing all aspects of quality system audits, from audit planning through following up on corrective action requests.
- ISO 10011-2: 1990, *Qualification criteria for quality systems auditors,* outlines educational, training, and personal and experiential requirements for quality system auditors and lead auditors.
- ISO 10011-3: 1990, *Management of audit programs,* provides important guidelines for organizing, staffing, and carrying out audits. While general enough to apply to all types of audits, its principles should be observed as facilities develop internal audit programs consistent with the ISO 9000 standard.

ISO 19011 defines auditing terms, sets forth the principal features of auditing, provides audit management guidelines, describes all aspects of auditing activities, and outlines auditor qualification and competence requirements. It will apply to both ISO 9000 and the ISO 14000 series of environmental management systems standards.

Guidelines for Measuring Equipment

A facility using an ISO 9000 quality system monitors critical quality characteristics on a planned basis to assess conformance to requirements. The standard does not specify how to do so. It is up to facility management to determine what is appropriate and to plan and conduct the operations accordingly.

For many facilities, however, monitoring means taking measurements. This requires the selection, confirmation, and periodic audit of measuring equipment, covered by ISO 9001: 1994, Element 4.11, and ISO 9001: 2000, Element 7.6. The effectiveness of measurements and decisions taken as a result is so critical that ISO also publishes ISO 10012-1: 1992, *Quality Assurance Requirements for Measuring Equipment—Part 1: Metrological confirmation for measuring equipment,* and ISO 10012-2: 1997, *Quality Assurance Requirements for Measuring Equipment—Part 2: Guidelines for control of measurement processes.*

ISO 10012-1 can be applied as agreed between individual facilities and their customers. It spells out the following requirements covering measuring equipment:

- Selection
- Confirmation
- Audit and review
- Measurement uncertainty
- Confirmation procedures
- Recordkeeping
- Handling nonconformances
- Storage and handling
- Traceability
- Personnel issues

ISO 10012-1 also includes implementation guidance, definitions, and a reference section.

Supporting Documents

A number of other documents provide insight, support, interpretation, and guidance to the principal ISO 9000 quality system documents mentioned earlier. These include

- ISO 9000-1: 1994, *Quality Management and Quality Assurance Standards—Part 1: Guidelines for selection and use,* provides a brief introduction to the standard. It includes definitions and guidelines on applying the standard for quality management and contractual purposes.

- ISO 8402: 1994, *Quality Management and Quality Assurance Vocabulary,* is a glossary of terms used in the various documents that describe the standard.
- ISO 9000-2: 1993, *Quality Management and Quality Assurance Standards—Part 2: Generic guidelines for the application of ISO 9001, 9002, and 9003,* contains generic guidelines for applying the three contractual models.

ISO 9000-1 and 8402 will be replaced in the 2000 revisions by ISO 9000: 2000, *Quality Management Systems—Fundamentals and vocabulary.*

ISO 9000 for Quality Management Purposes: The Guidance Documents

ISO 9004-1, 9004-2, 9004-3, and 9004-4 are parts of the standard used by facilities that wish to implement quality systems but have no contractual obligation to do so. These documents will be replaced under the 2000 revisions by ISO 9004, *Quality Management Systems—Guidelines for performance improvement,* whose structure parallels that of ISO 9001: 2000. This said, it also should be stressed that facilities seeking registration to ISO 9001, 9002, or 9003 would be wise to consult these guidance documents as well.

Arguably, ISO 9004: 2000, 9004-1, and 9004-2 provide a clearer and more graphic illustration of an integrated quality system than do the contractual models. ISO 9004-3 is a specialized set of guidelines for processed materials. ISO 9004-4 sets guidelines for quality improvement.

All these documents provide amplification, clarification, and additional guidance on the topics covered in the contractual models. In addition, they address certain issues that are not covered in the contractual models but which are important to the effective operation of quality systems. Examples are quality costs and product safety and liability.

ISO 9004-1, and its more generic successor ISO 9004: 2000, can and should be consulted by both manufacturing and service-oriented facilities. ISO 9004-2 is written specifically for organizations in the service sector. Most manufacturing firms, however, have at least some service element within their scope of operations, so it is wise for managers to be familiar with both sets of guidelines, pending the publication of ISO 9004: 2000, which applies to both manufacturing and service facilities.

ISO 9004-1, ISO 9004-2, and ISO 9004: 2000 are constructed around a graphic device called a *quality loop,* which is also used for the process-based structure of ISO 9001: 2000. We will look closely at quality loops in Chap. 5. For now, it is sufficient to say that the quality loop illustrates the major components of the quality system in circular form, beginning and ending with customers.

ISO 9004: 2000 sets forth, in Section 4.3, eight quality management principles that have been identified to facilitate the achievement of quality objectives. These principles are the basis of ISO 9001: 2000 and also appear in ISO 9000: 2000, Section 0.2. They are

- *Customer focus.* Organizations depend on their customers and therefore should understand current and future customer needs, meet customer requirements, and strive to exceed customer expectations.
- *Leadership.* Leaders establish unity of purpose, direction, and the internal environment of the organization. They create the environment in which people can become fully involved in achieving the organization's objectives.
- *Involvement of people.* People at all levels are the essence of an organization, and their full involvement enables their abilities to be used for the organization's benefit.
- *Process approach.* A desired result is achieved more efficiently when related resources and activities are managed as a process.
- *Systems approach to management.* Identifying, understanding, and managing a system of interrelated processes for a given objective contributes to the effectiveness and efficiency of the organization.
- *Continual improvement.* A permanent objective of the organization is continual improvement.
- *Factual approach to decision making.* Effective decisions are based on the logical or intuitive analysis of data and information.
- *Mutually beneficial supplier relationships.* The ability of the organization and its suppliers to create value is enhanced by mutually beneficial relationships.

ISO 9004-1, 9004-2, and 9004: 2000 provide guidelines for the development and application of quality systems (see Table 4-1).

In Part 2 of this book we examine the specific technical requirements and guidelines of the ISO 9000 standard, including the recommendations in ISO 9004-1, ISO 9004-2, and ISO 9004: 2000. Keep in mind that these documents contain no requirements per se, only recommendations.

ISO 9000 for Contractual Purposes: The Quality System Models

The contractual quality system models, ISO 9001, 9002, and 9003; contain the ISO 9000 requirements for facilities seeking to become

TABLE 4-1. ISO 9004 Cross-References

ISO 9004-1: 1994	ISO 9004-2: 1991	ISO 9004: 2000
0 Introduction		0 Introduction
1 Scope	1 Scope	0.1 General
2 Normative references	2 Normative references	0.2 Process approach
3 Definitions	3 Definitions	0.3 Relationship with ISO 9001
4 Management responsibility	4 Characteristics of services	0.4 Compatibility with other management systems
5 Quality system elements	4.1 Service and service delivery characteristics	1 Scope
6 Financial considerations of quality systems	4.2 Control of service and service delivery characteristics	2 Normative reference
7 Quality in marketing	5 Quality system principles	3 Terms and definitions
8 Quality in specification and design	5.1 Key aspects of a quality system	4 Quality management system guidelines
9 Quality in purchasing	5.2 Management responsibility	4.1 Managing systems and processes
10 Quality of processes	5.3 Personnel and material resources	4.2 General documentation requirements
11 Control of processes	5.4 Quality system structure	4.3 Use of quality management principles
12 Product verification	5.5 Interface with customers	5 Management responsibility
13 Control of inspection, measuring, and test equipment	6 Quality system operational elements	5.1 General guidance
14 Control of nonconforming product	6.1 Marketing process	5.2 Needs and expectations of interested parties
15 Corrective action	6.2 Design process	5.3 Quality policy
16 Postproduction activities	6.3 Service delivery process	5.4 Planning
17 Quality records	6.4 Service performance analysis and improvement	5.5 Administration
18 Personnel		5.6 Management review
19 Product safety		6 Resource management
20 Use of statistical methods		6.1 General guidance
		6.2 People
		6.3 Infrastructure
		6.4 Work environment

TABLE 4-1. ISO 9004 Cross-References (*Continued*)

ISO 9004-1: 1994	ISO 9004-2: 1991	ISO 9004: 2000
		6.5 Information
		6.6 Suppliers and partnerships
		6.7 Natural resources
		6.8 Finance
		7 Product realization
		7.1 General guidance
		7.2 Processes related to interested parties
		7.3 Design and/or development
		7.4 Purchasing
		7.5 Production and service operations
		7.6 Control of measuring and monitoring devices
		8 Measurement, analysis, and improvement
		8.1 General guidance
		8.2 Measurement and monitoring
		8.3 Control of nonconformance
		8.4 Analysis of data for improvement
		8.5 Improvement

registered to the standard (see Table 4-2). Facilities become registered to the part of the standard required by the contract and/or that most closely reflects the scope of the facility's process. Under the 2000 revisions, facilities will only register to ISO 9001, which could be tailored to fit the scope of the facility's process.

ISO 9001 is the most comprehensive part of the 1994 standard. It covers facilities whose process includes design, development, production, installation, and servicing. It contains 20 quality system elements.

The scope of ISO 9002 is more limited. It applies to facilities making products that are designed and serviced by customers or subcontractors. ISO 9002 includes 19 of the 20 ISO 9001 elements, excluding the design control element (4.4).

ISO 9003 is the most limited part of the standard. It applies only to facilities performing final inspection and test functions. It incorporates 16 of the 20 ISO 9001 elements, excluding Elements 4.4

TABLE 4-2. ISO 9001 Cross-References

ISO 9001: 1994	ISO 9001: 2000
4.1 Management responsibility	5.1, 5.3, 5.4.1, 5.5.2, 5.5.3, 5.6, 6.1, 6.2.1, 6.3
4.2 Quality system	4, 5.1, 5.4.1, 5.4.2, 5.5.5, 7.1
4.3 Contract review	7.2.2
4.4 Design control	7.3
4.5 Document and data control	5.5.6
4.6 Purchasing	7.4
4.7 Control of customer-supplied product	7.5.3
4.8 Product identification and traceability	7.5.2
4.9 Process control	7.1, 7.5.1, 7.5.5
4.10 Inspection and testing	7.1, 7.5.1, 8.1, 8.2.4
4.11 Control of inspection, measuring, and test equipment	7.6
4.12 Inspection and test status	7.5.1
4.13 Control of nonconforming product	8.3
4.14 Corrective and preventive action	8.4, 8.5.2, 8.5.3
4.15 Handling, storage, packaging, preservation, and delivery	7.1, 7.5.4
4.16 Control of quality records	5.5.7
4.17 Internal quality audits	8.2.2
4.18 Training	6.2.2
4.19 Servicing	7.1, 7.5.1
4.20 Statistical techniques	8.1, 8.2.3, 8.2.4, 8.4

(design control), 4.6 (purchasing), 4.9 (process control), and 4.19 (servicing).

Aside from the differences among the parts, these models were meant to be generic. They apply to virtually any facility producing any product or service for virtually any market. Keeping the ISO standards as generic as possible maximizes ease of translation and applicability throughout the industrialized world.

In meeting changing marketplace needs, ISO 9001: 2000 is designed to be more generic, follow a process-based structure and eight quality management principles, and be more compatible with the ISO 14000 series of environmental management systems (EMS) standards, allowing for integrated management systems. It allows the new option of being tailored to omit requirements that do not apply to an organiza-

TABLE 4-3. ISO 9001: 2000 Elements

ISO 9001: 2000	ISO 9001: 1994
4 Quality management system	
4.1 General requirements	4.2.1
4.2 General documentation requirements	4.2.2
5 Management responsibility	
5.1 Management commitment	4.1, 4.2.1
5.2 Customer focus	
5.3 Quality policy	4.1.1
5.4 Planning	4.1.1, 4.2.1, 4.2.3
5.5 Administration	4.1.2, 4.2.1, 4.5, 4.16
5.6 Management review	4.1.3
6 Resource management	
6.1 Provision of resources	4.1.2.2
6.2 Human resources	4.1.2.1, 4.18
6.3 Facilities	4.9
6.4 Work environment	4.9
7 Product realization	
7.1 Planning of realization process	4.2.3, 4.9, 4.10, 4.15, 4.19
7.2 Customer-related processes	4.3
7.3 Design and/or development	4.4
7.4 Purchasing	4.6
7.5 Production and service operations	4.7, 4.8, 4.9, 4.10, 4.12, 4.15, 4.19
7.6 Control of measuring and monitoring devices	4.11
8 Measurement, analysis, and improvement	
8.1 Planning	4.10, 4.20
8.2 Measurement and monitoring	4.10, 4.17, 4.20
8.3 Control of nonconformity	4.13
8.4 Analysis of data	4.14, 4.20
8.5 Improvement	4.1.3, 4.9, 4.14

tion or limit the scope of application under Section 1.2, Permissible Exclusions. This eliminates the need for the less comprehensive ISO 9002 and 9003 standards. In addition, there are no QMS documentation layout or structure requirements.

The new process-based structure, similar to that used in the ISO 14001 EMS standard, creates a completely different look for ISO 9001. The 20 elements will be replaced by five clauses containing 23 elements. ISO 9004: 2000 is organized identically (see Table 4-3). ISO 9001 and 14001 are more compatible under this approach, making it easier to integrate management systems and combine documentation.

As you examine the requirements and guidelines of ISO 9000, you will find nothing radical or new. The standard is, in fact, a commonsense mix of sound business practices and time-proven quality methods. The focus is not on products or services. It is on the facility's quality system, the network of activities designed and operated to ensure that output meets the ultimate business objective of satisfying the customer.

The ISO 9000 standards stress objectives, not methods; concepts, not procedures. Finally, the standard gives no criteria whatever for product or service features such as hardness, durability, or response time. Instead, the requirements and guidelines of the ISO 9000 standard are an intentionally bare-bones blueprint of the ideal quality system. Next, in Part 2, we will take a detailed look at that blueprint.

Part

2

Technical Requirements and Guidelines

As discussed in Chap. 4, the heart of ISO 9000 is a series of models and several sets of guidelines, which will be reduced to one model and one set of guidelines under the 2000 revisions. Each applies to facilities engaged in a specific scope of activities, using ISO 9000 for a specific purpose, whereas the 2000 drafts are more generic.

- Facilities wishing to register to ISO 9000 adhere to one of the contractual models, ISO 9001, 9002, or 9003, depending on their scope of operation. Under the 2000 revisions, facilities will only register to ISO 9001, which could be tailored to fit their scope of operation under Section 1.2, Permissible Exclusions.
- Facilities engaged in manufacturing that do not seek registration but want to implement a quality system consistent with the standard rely on ISO 9004-1: 1994 for guidance.
- Facilities engaged in service that do not seek registration but want to implement a quality system consistent with the standard rely on ISO 9004-2: 1991.
- ISO 9004: 2000 will apply to both manufacturing and service facilities that do not seek registration but want to implement a quality system consistent with ISO 9001: 2000. It also should be consulted for guidance by facilities seeking registration.

Confusion has occurred because the 20 elements of ISO 9001, while covering much the same subject matter as sections of the ISO 9004-1 and 9004-2 guidance documents, are in a completely different order. Under the 2000 revisions, on the other hand, ISO 9001 and 9004 are organized identically, thereby eliminating this confusion.

In developing an ISO 9000 quality system, many facilities find it necessary to draw on more than one of the models or guidelines, a problem that will not occur with the 2000 revisions. This is so because many facilities are engaged in both manufacture and service in varying proportions. A maker of automotive subassemblies, for example, is almost entirely involved in manufacture and only slightly involved in service. Restaurants are a virtual 50-50 mix between the two. Law offices provide 90 percent service and only 10 percent, or less, product.

In order to design an appropriate ISO 9000 quality system, managers should become familiar with all technical requirements and recommendations. The full array of technical requirements and recommendations is explored in the following chapters.

For coherence, and to eliminate the confusion and redundancy inherent among the various old ISO 9000 documents, the requirements have been organized and reordered by subject. Appendix A provides a detailed cross-reference, by subject and section number(s), among the various old ISO 9000 models and guidelines. The 2000 revisions, by reducing the number of documents and organizing them identically, avoids this confusion.

Because the standard itself is worded fairly generally, the presentation of the requirements is general too. The applicability of any particular passage depends on the nature and scope of the facility in question, a matter recognized in ISO 9001: 2000, which can be tailored appropriately. Many requirements apply to all facilities. Others apply in varying degrees, whereas still others do not apply to certain facilities at all. For example, facilities seeking registration to ISO 9002 will have no use for the design or service elements.

The following chapters are intended to provide a reliable guide to the technical requirements and recommendations of ISO 9000 and the principles behind them. It is up to management to apply these requirements to its own facility and process to determine exactly what the particular ISO 9000 quality system needs.

Appendix B presents a self-assessment checklist to aid facility managers in this process. For definitive help, the guidance of an experienced ISO 9000 consultant or registrar is strongly recommended. Because of their expertise, consultants can implement an ISO 9000 quality system with much greater efficiency than facility management acting alone.

One final word: Not every reader has the need or desire to become immersed in the minutiae of the standard. For this reason, each element begins with a brief summary in checklist form of *requirements* and *recommendations* of the element under discussion. The text that follows each summary provides details, explanations, and comments for readers who want more in-depth coverage of the material.

Finally, all readers are encouraged to read the standards themselves, particularly the 2000 revisions when they take effect in November or December.

Glossary of Terms

Customer The recipient of a supplier's output, which may be internal as well as external.

Output A product or service.

Subcontractor An entity providing output for the use of a supplier or facility.

Supplier or facility A product or service provider that is adopting an ISO 9000 quality system.

Chapter

5

Developing and Managing the Quality System

What I expect to have with ISO 9001 is a finely tuned quality system which gives us a smooth flow of information throughout the product cycle, from cradle to grave.

JIM ECKLEIN
Augustine Medical
Registered to ISO 9001, EN 46001 and the EU Medical Device Directive

ISO 8402, the quality management vocabulary, defines a *quality system* as "the organizational structure, responsibilities, procedures, processes and resources needed to implement quality management." ISO 9000: 2000, Element 2.2.3, defines a *quality management system* as a "system to establish a quality policy and quality objectives and to achieve those objectives." Every element of the standard concerns the quality management system, of course. But several elements specify and define the basic components:

Management responsibility

Quality system

Quality costs

Internal quality audits

Personnel and training

These requirements are integral to the quality system and affect virtually every kind of facility that adheres to the standard. They are examined in detail below.

Management Responsibility (ISO 9001: 1994, 4.1; ISO 9004-2: 1991, 5.2; ISO 9001 and 9004: 2000, 5.1, 5.3, 5.4.1, 5.5.2, 5.5.3, 5.6, 6.1, 6.2.1, 6.3)

Checklist of requirements:

- Top management designates a representative with authority and responsibility for implementing and maintaining the requirements of the standard and reporting on quality system performance to management.
- Top management establishes, documents, and publicizes its policy, objectives, and commitment to quality and customer satisfaction.
- Top management defines the responsibility, authority, and relationships for all employees whose work affects quality.
- Top management conducts in-house verification and review of the quality system.
- The organization identifies and provides resource requirements necessary to ensure the proper functioning of the quality system.

Every quality guru from Deming on has asserted that the success of a quality system is directly related to the consistency and intensity of top management's commitment. The requirements and guidelines of ISO 9000 acknowledge the truth of this claim.

The *management responsibility* requirements, which come first in the standard, place the burden directly on top management, where it belongs. ISO 9004-1 states it most bluntly in Element 4.1: "The responsibility for and commitment to a quality policy belongs to the highest level of management." This is reiterated in ISO 9001: 2000, Element 5.4.1, which states, "Top management shall ensure that quality objectives are established at relevant functions and levels within the organization."

In sum, the standard obligates top management to define its quality policy and execute it through an organization of people and resources. Management is also obliged to participate actively in the quality system by conducting verification and review activities.

First of all, the standard requires top management to designate a *management representative* (MR). This individual bears day-to-day responsibility for the quality system and for adherence to the standard. The MR should possess sufficient rank and authority within the facility to be able to develop, monitor, and change elements of the quality system. The MR's other responsibilities usually include liaison with the registrar and oversight of the facility's internal audit program.

The standard also requires top management to create and document a quality policy. The policy should address acceptable quality levels, the facility's quality image and reputation, quality objectives, a strategy for achieving those objectives, and the responsibility of employees for executing the strategy. The policy must be relevant to the organization and must relate to the needs and requirements of customers.

ISO tacitly recognizes that in many firms quality policies are formulated, often after much roaring and screaming, and then promptly filed away and forgotten. The standard does not permit this to happen, at least the filed and forgotten part. It requires that the quality policy be documented and publicized so that it is understood at all levels of the facility. The facility must be able to show how awareness is achieved on an ongoing basis with new and existing employees.

Under the management responsibility element, top management is also required to define and document responsibility, authority, and relationships for all employees whose work affects quality. This does not necessarily result in lengthy and formal job descriptions. Job descriptions are one option, whereas work instructions are another.

The objective set by this section is eminently reasonable to any responsible manager: to organize the facility in such a way that the work is carried out by people who understand what to do, how to do it, and how to deal with one another. As with so many other requirements of the standard, this one specifies the ends, not the means.

The standard further acknowledges that change is a constant and that quality systems are never perfect. Management must pay consistent attention to the quality system to ensure that it improves and adjusts to change. Accordingly, the standard calls on management to conduct regular *verification* of the quality system and the processes that it governs. Verification activities must be carried out only by people who do not have direct responsibility for the activities being verified. One form of verification is the *internal audit,* which we will examine later in this chapter.

In addition, the MR and other members of management are responsible for conducting regular, documented *management reviews* of the quality system at defined intervals. These reviews consider, at a minimum:

- Results of internal quality audits
- Management effectiveness
- Defects and irregularities
- Resolution of customer complaints
- Solutions to quality problems

- Implementation of corrective actions
- Handling of nonconforming product
- Results of statistical scorekeeping tools
- Impact of quality methods on actual results

Reviews must, of course, be fully documented. Management must show that it uses the results of these reviews as a basis for improving the quality system in particular and the facility's quality in general.

As befits the section that comes first, management responsibility is arguably the most important component of the standard. This confirms what quality professionals have always known: that management responsibility and commitment are absolute prerequisites to the successful pursuit of quality.

New management responsibility requirements in ISO 9001: 2000 include establishment of quality objectives (5.4.1), determining customer needs and expectations (5.2), identifying and planning resources needed to achieve quality objectives (5.4.2), the MR promoting awareness of customer requirements (5.5.3.c), an expanded quality policy (5.3), management review input (5.6.2), and management review output (5.6.3).

Quality System (ISO 9001: 1994, 4.2; ISO 9004-2: 1991, 5.3.3, 5.4.1-2; ISO 9001 and 9004: 2000, 4, 5.1, 5.4.1, 5.4.2, 5.5.5, 7.1)

Checklist of requirements:

- A quality system shall be established, documented, and maintained to ensure that products or services meet specified requirements.
- The quality system shall be documented in a quality manual or equivalent document and make reference to quality system procedures and ISO 9000 requirements.
- Requirements for achieving quality objectives shall be defined and documented.
- A system for developing and implementing quality plans should be considered for products, services, projects, or contracts. This system should consider

  Identification and acquisition of adequate resources, such as controls, processes, equipment, and skills, to meet quality objectives, with due consideration given to required levels of experience, training, and general competence levels of personnel.

  Preparation of quality plans.

  Identification of appropriate verification activities at appropriate production stages.

Identification and preparation of quality records.

Defined standards of acceptability for all features and requirements.

Methods for updating quality control and inspection and testing techniques as necessary.

Processes for ensuring the compatibility of the design, installation, servicing, inspection, and test procedures, production process, and relevant documentation.

Most facilities have quality systems of one form or another. Some are intricate and extensive. Others are rudimentary. Although quality systems are what ISO 9000 is all about, the standard does not oblige facilities to add on a separate and sometimes redundant structure. Instead, the standard

- Describes the *ideal* quality system in illustrative form
- Sets forth *common-sense goals* for an adequate or acceptable quality system
- Specifies the *minimum essentials* a quality system should include to achieve those goals

Managers of many facilities are pleasantly surprised to discover that their existing quality systems already meet ISO 9000 requirements in many respects. There is, after all, nothing new or radically different in the standard. And the standard emphasizes goals and objectives, as outlined earlier. It does not prescribe specific strategies, tactics, and procedures. This is what gives ISO 9000 its enormous flexibility.

The quality loop

The standard describes the ideal quality system in graphic form as a quality loop. One version of the loop is for manufacturing (ISO 9004-1; Fig. 5.1), another is for service (ISO 9004-2; Fig. 5.2), and ISO 9001: 2000 uses a quality management process model (Fig. 5.3). The quality loop includes all stages of the quality system and shows their relationships. Notice that the gathering of customer requirements is a prominent part of these continuous loops. ISO 9004: 2000 uses the same quality management process model as ISO 9001: 2000, expanding from customers to interested parties, described in Element 5.2, to also include end users, people in the organization, owners and investors, suppliers and partners, and society.

The technical requirements and recommendations in this section, and indeed, throughout this book, all fit neatly into the quality-loop schematic, creating an effective and dynamic quality system.

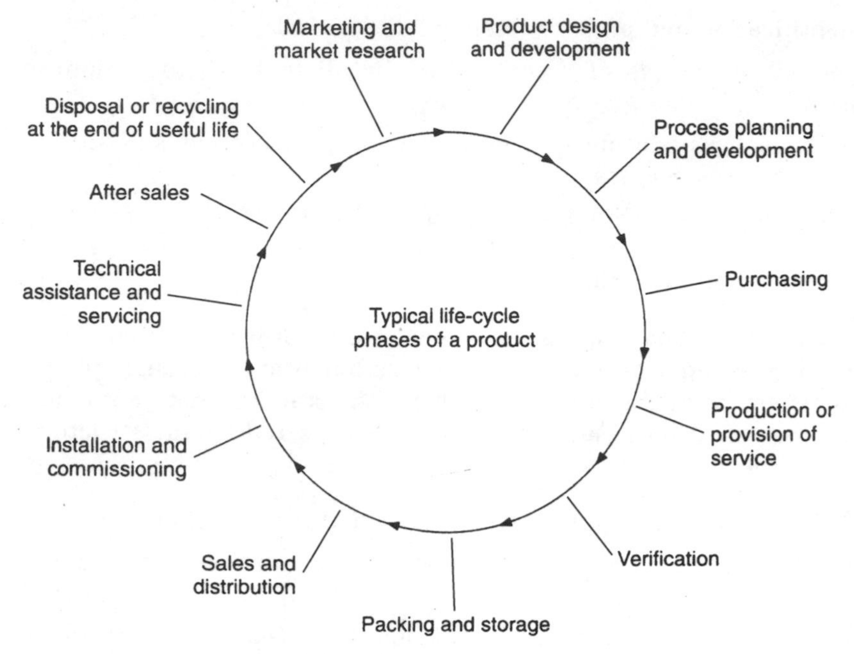

Figure 5-1. Quaility loop for ISO 9004-1.

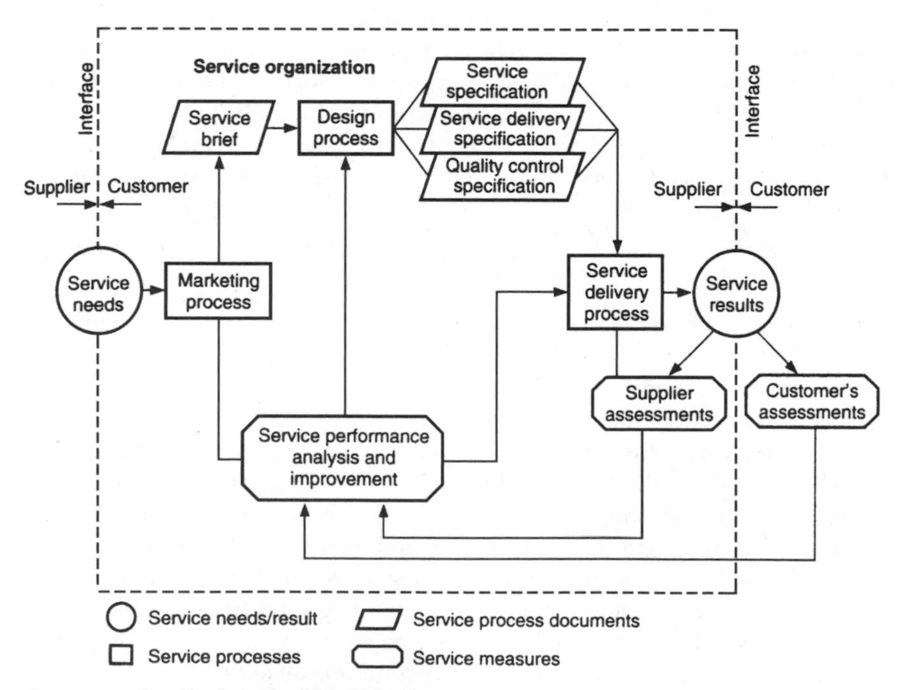

Figure 5-2. Quality loop for ISO 9004-2.

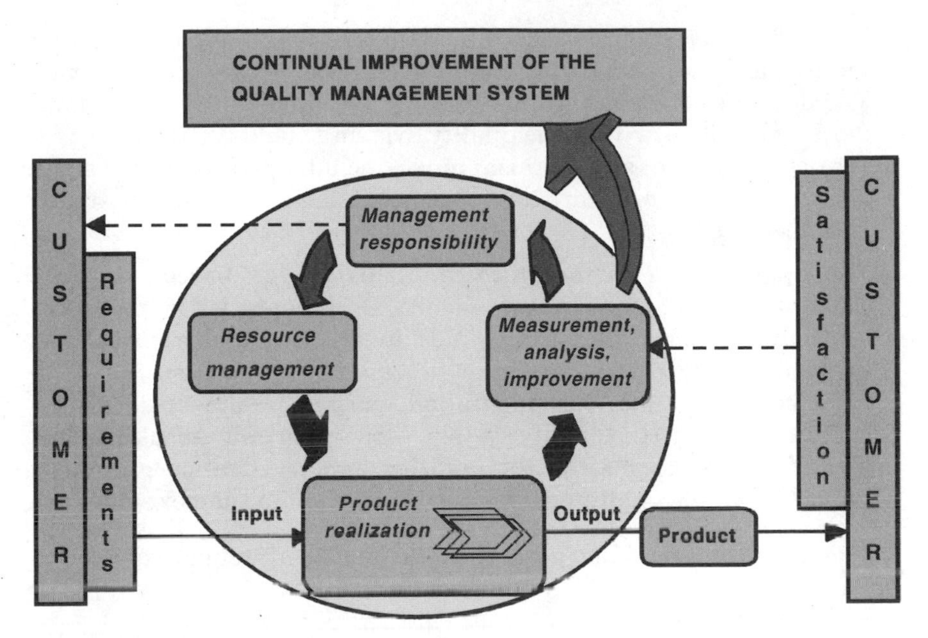

Figure 5-3. Quailty management process model.

Quality system goals

As noted earlier, the standard's principal intent is to prescribe goals and objectives, not specific tactics. Here, according to the standard, are the goals of the ideal quality system:

The system must be well understood and effective. Understanding is a function of documentation and training, both covered elsewhere in the standard. Effectiveness is self-explanatory, and means for enforcing audits, corrective action, reviews, and other requirements are also built into the standard, as we shall see.

The system must be effective at achieving quality objectives. This is self-explanatory, and it follows logically from the requirements under management responsibility. Why have a quality system if it does not meet the facility's quality goals?

The system must provide confidence that output meets customer expectations. This is usually an automatic quality objective, but since it is the whole purpose of having a quality system, it is expressed as a separate goal.

The system should give emphasis to preventive actions. At the same time, it should maintain the ability to detect, respond to, and correct failures after they occur. Some managers, pointing to the standard's sections on inspection and related activities, charge that the standard is based on detection. In fact, the standard's true goal, a higher and more logical one, is that *no nonconforming product reach the customer.* The first line of defense against this is prevention, and here the standard makes it clear that the quality system should be based on prevention. But the system also must include detection measures adequate enough to ensure that nonconforming output does not reach customers.

The system must be documented. This too is based on logic. Turnover is a fact of life in every facility. Documentation helps ensure continuity. Documentation also aids in training and ensures that everyone involved in a particular process is singing from the same page in the hymnal. Documentation exposes weaknesses, trouble spots, and gaps in the quality system. More practically speaking, documentation is essential to the auditing and registration processes. We will examine quality system documentation in the next chapter.

These are the overriding goals that the quality system is expected to achieve.

Quality system elements

The ISO 9000 standard and its various models and guidelines are, in fact, blueprints for what ISO considers to be the ideal quality system. Having specified some generic, rather obvious, and quite logical goals for the quality system, the standard goes on to set forth a roster of elements that it considers important for an effective quality system to achieve the ultimate business goal of producing output that meets customer requirements.

Among these important elements is a quality manual, required by ISO 9001: 1994, Element 4.2.1, and ISO 9001: 2000, Element 5.5.5. The standard requires a quality manual that either includes or refers to procedures that define the quality system. These procedures, in turn, must be prepared and documented according to the requirements of both the ISO 9000 standard and the company's quality policy.

Another important element of the quality system is quality planning, required by ISO 9001: 1994, Element 4.2.3, and ISO 9001: 2000, Element 5.4.2. ISO 9000 requires quality planning that takes into account several important contract and product specification considerations, including quality plans, equipment, resources, production processes, verification activities, quality records, and acceptability standards. Under ISO 9001: 2000, quality planning must include continual improvement of the quality management system.

It is up to facility management, assisted if necessary by experienced ISO consultants and auditors, to determine the precise shape of the quality system appropriate to the facility's business, marketplace, and objectives. You will find that the ISO 9000 standard is flexible enough to accommodate virtually all these parameters.

Quality Costs (ISO 9004-1: 1994, 6; ISO 9004: 2000, 6.8, 8.2.1.4)

Checklist of requirements:

- Operating quality costs and external assurance quality costs should be identified and measured.
- Quality costs should be reported to and monitored by management in the context of other costs and revenues.

Quality in general has a profound effect on the facility's financial health or lack thereof. The ultimate purpose of any quality program is to maintain and improve the facility's health by satisfying its customers. A valuable way to measure the effectiveness of a quality system is to measure and monitor its costs, as well as the return on investment that it generates. For this reason, the standard recommends that two broad categories of quality costs be monitored, analyzed, and reported to management.

Types of quality costs

According to ISO 9004: 2000, Element 6.8, *internal* quality costs include process and product failures, or waste in material and time, whereas *external* quality costs include product failures, costs of compensation of guarantees and warranties, and costs of lost customers and markets.

Cost analysis approaches

ISO 9004-1 offers three approaches for gathering, presenting, and analyzing financial data. Each can be useful for determining and measuring the financial impact of implementing an effective quality system.

- *Quality cost approach.* Quality-related costs arising from internal and external activities are identified and measured. These costs, in turn, are broken down into four categories: *prevention,* efforts undertaken to prevent failures; *appraisal,* testing, inspection, and examination to determine whether the requirements of the standard are being met; *internal failure,* costs associated with a product or service failing to meet quality requirements prior to delivery; and *external failure,* costs resulting from a product or service failing to meet quality requirements after delivery. Costs associated with prevention and appraisal are considered *investments,* whereas failure costs are considered *losses.*

- *Process cost approach.* Costs of *conformity,* the expense of meeting all standard requirements and customer needs, and costs of *nonconformity,* which result from a failing process, are defined and analyzed.
- *Quality loss approach.* Internal and external losses resulting from poor quality are identified and categorized as tangible and intangible. Loss of future sales due to customer dissatisfaction is an *external* intangible loss. Lowered work efficiency due to rework is an *internal* tangible loss.

ISO 9004: 2000, Element 8.2.1.4, makes no financial approach recommendations. Instead, it presents three examples: prevention, appraisal, and failure costs analysis; costs of conformance and nonconformance; and life cycle.

Quality cost reporting

ISO 9004: 2000 recommends that quality cost information be provided for management review as part of continual improvement. Clearly, the standard regards quality cost reporting as an important element of management planning. Suggested uses of quality cost reports include

- Evaluating quality system adequacy and effectiveness.
- Pinpointing areas requiring improvement.
- Establishing quality and cost objectives.

Internal Quality Audits (ISO 9001: 1994, 4.17; ISO 9004-1, 5.4; ISO 9004-2, 5.4.4; ISO 9001: 2000, 8.2.2; ISO 9004: 2000, 8.2.1.3)

Checklist of requirements:

- Regular internal audits must be conducted to evaluate how effectively a quality system is achieving stated objectives and conforming to ISO 9000.
- Audits must be conducted on a regular basis in accordance with a documented audit plan.
- An audit program should be established and implemented by management. The program should cover audit planning and scheduling, the assignment of personnel to conduct audits, and documented procedures for performing audits, including recording and reporting audit findings.

- Audit findings must be documented and acted on by top management in the form of corrective actions to eliminate nonconformances and verification of corrective action implementation.

Audits are a major element in an ISO 9000 quality system. Under ISO, there are two types of regular external audits, as well as a requirement for internal auditing measures.

Facilities registered to one of the contractual models, ISO 9001, 9002, or 9003, undergo initial audits by the registrar. Registered facilities are also subject to surveillance audits, during which various aspects of the quality system are examined. Typically, all aspects of an organization's quality system must be examined within 3 years. We will delve into these types of audits in Chap. 13.

The standard also obligates management to take regular measures to audit the quality system and ensure that it is meeting stated objectives, as required under management responsibility. To meet this requirement, the standard requires facilities to conduct regular internal audits to verify the suitability and effectiveness of the quality system.

The audit plan

The facility must conduct these internal audits in accordance with a documented plan, which should cover every element of the quality system. The standard suggests that areas for internal audit attention should include

- Procedures (documented versus actual)
- Resources
- Facilities
- Measurement systems and results
- Documentation and records
- Nonconformances reported in previous audits

The following factors should be spelled out in the facility's audit plan:

- Areas and activities to be audited.
- Frequency of audits. The standard does not mandate frequency. It is up to management to determine when an internal audit is appropriate. In practice, it is prudent for every element of the quality system to undergo an internal audit at least once a year.
- Qualifications of persons carrying out audits. Again, these are up to the judgment of facility management. The standard requires that auditors have no involvement in the areas being audited.

- Basis for carrying out audits. In addition to timing issues, the plan may call for audits to be triggered by such events as a certain level of customer complaints, suspect statistical analyses, or excessive nonconformances detected in previous internal audits.
- Procedures for reporting audit findings and recommendations.

Audit findings

Audit results should be documented and brought to the attention of the management of audited areas. The standard defines results to include nonconformances, suggested corrective actions, and follow-up actions to verify the effectiveness of corrective actions arising from earlier audits.

Personnel and Training (ISO 9001: 1994, 4.18; ISO 9004-1, 18; ISO 9004-2, 5.3; ISO 9001 and 9004: 2000, 6.2)

Checklist of requirements:

- Employees assigned to work affecting quality shall meet established qualification standards based on education, training, and/or experience.
- The facility must have a documented system that identifies training needs of people whose work affects quality.
- The facility should provide such training where needs exist.
- The facility must maintain training records.
- The facility should strive to motivate all employees toward quality performance.
- The need for quality should be stressed through an awareness program.
- An objective measurement system should be developed to make employees aware of their achievements. Performance also should be recognized.

The ISO 9000 standard recognizes how vital *people* are to the success of a quality system. Accordingly, the standard sets forth detailed requirements and guidelines governing qualification, training, and motivation of employees whose work affects quality in any way.

Interestingly, the requirements in the ISO 9001, 9002, and 9003 contractual models are rather sparse. It is the ISO 9004-1 guidelines document that spells out the details. ISO 9001: 2000 offers more details than its predecessor, with further elaboration in ISO 9004: 2000.

Employee qualifications

Some jobs, especially technical ones, are best filled by employees meeting certain formal qualifications. These qualifications may be in the form of certifications, licenses, or demonstration of special skills. The standard advises facilities to have a system for determining what positions fall into this category. In such cases, the facility should implement procedures for verifying that necessary qualifications have been met.

Who should undergo training?

The ISO 9000 standard does not in itself require training of any kind. This is, once again, an area left up to the best judgment of facility management. The standard acknowledges, for example, that many employees may meet the qualifications for their positions by virtue of experience.

However, the standard clearly regards training as a good thing per se. It suggests that facility management consider some form of training for everyone. It makes special mention of people whose work affects quality, of course, as well as new hires and transferees, as logical candidates for training.

What kind of training should be provided?

In general, facilities should provide training to employees in accordance with their documented training plan. They also should provide whatever training is required to ensure that employees meet the documented qualifications of their positions. The standard makes more specific suggestions for various employee levels:

- *Management personnel.* Quality system understanding
- *Technical personnel.* Technical skills needed to meet quality plan obligations, as well as statistical techniques
- *Line management and hourly workers.* Skills needed to perform assigned tasks and meet quality plan obligations, safety, and (potentially) statistical techniques

What kinds of motivation methods are required?

In a word, none. This should come as no surprise. The standard does acknowledge that motivational techniques are commonly used in facilities today. It counsels facilities to use motivational methods that

- Foster worker involvement
- Emphasize the role of quality in accomplishing tasks

- Emphasize the impact of quality on the facility's success
- Recognize and reward efforts that contribute to success of the quality system

This chapter has outlined the ISO 9000 requirements and guidelines with regard to some basic elements of the facility's quality system. These are very general areas that apply to virtually every quality system. In the next few chapters we will look at what the standard has to say about some finer points of the quality system. We will see what it says about quality in subcontractor and customer relationships, and we will explore everyone's favorite topic—documentation.

Chapter

6

Documenting the Quality System

ISO 9000 is a paperwork nightmare and record storage nightmare.

ISO 9000's high structure and documented system will stifle innovation and initiative—paperwork over all else!

RECENT SEMINAR ATTENDEES

Documentation is central to ISO 9000. And nothing else in ISO 9000 brings out management fear and loathing as strongly as does its emphasis on documentation. Many firms see ISO 9000 as unnecessarily paper-heavy and bureaucratic. Some go so far as to say that ISO 9000 is not about quality at all but about redundant recordkeeping and paper shuffling.

At first blush, it is easy to understand the concern some have about ISO 9000's documentation requirements. The standard itself is rife with references to documentation. ISO 9001: 1994 devotes two elements to it and refers to documentation and recordkeeping in no fewer than 50 other places. ISO 9004-1 also devotes two sections to documentation requirements. ISO 9004-2 describes the quality system as consisting of three main elements: the quality loop, internal quality audits, and documentation. The more streamlined ISO 9001 and 9004: 2000 devote one element to documentation, but with no less detail than their predecessors.

Is the standard's emphasis on documentation and recordkeeping excessive, unnecessary, and burdensome? To answer this question, let us look at the objectives for documentation as set forth in the standard:

> The quality system should require that sufficient records be maintained to demonstrate conformance to specified requirements and verify effective operation of the quality system.
>
> *ISO 9004-1: 1994, 17.2*

ISO 9004: 2000, Element 5.5.5, contains virtually identical language. Does this sound onerous or threatening? No. In fact, it sounds rather reasonable. Certainly, as any businessperson can attest, some form of documentation and recordkeeping is essential to operating any business.

Consider another statement in the standard:

> Care should be taken to limit documentation to the extent pertinent to the application.
>
> *ISO 9004-1: 1994, 5.3.1*

Similar language appears in ISO 9004: 2000, Element 4.2. ISO 9000, then, *requires* only a reasonable and prudent level of recordkeeping and documentation and even cautions the facility against overdoing it. Therefore, the facility adopting ISO 9000 need not create redundant recordkeeping or documentation systems. Instead, as we will see in Chap. 12, the facility should approach the matter this way:

- Compare the standard's requirements against its existing operation.
- Adapt existing documentation and recordkeeping systems as necessary to conform to the standard.
- Then, and only then, eliminate any remaining nonconformances with additional documentation and recordkeeping.

The various parts of ISO 9000 describe documentation requirements in several different sections, as examined in detail below:

Documentation of the System

Document and Data Control

Quality Records

Documentation of the Quality System (ISO 9001: 1994, 4.2.1; ISO 9004-1, 5.3; ISO 9004-2, 5.4.3.1; ISO 9001 and 9004: 2000, 4.2, 5.5)

Checklist of requirements:

- The quality system should be documented in a systematic and orderly manner in the form of written policies and procedures.
- Quality system documentation should be understood, agreed to, and accessible to all personnel whose activities affect quality.

The ISO 9000 quality system is intended to be planned, orderly, controlled, and verifiable. Policies and procedures are necessary to ensure that

- Employees know what to do.
- Management maintains control.
- Oversight bodies, such as registrars, can attest to the effectiveness of the quality system.

Quality system documentation is usually composed of several different levels. Each of these levels is defined on the following pages, beginning with the quality manual.

The quality manual

The top tier of quality system documentation is usually a *quality manual.* Of all the elements that make up the ISO 9000 quality system, none is as central and important as the quality manual. It serves a multitude of essential purposes. The quality manual

- Aids in creating and implementing the quality system
- Describes the objectives and structure of the quality system
- Demonstrates management's commitment to the quality system
- Serves as a cross-reference between the quality system and the quality system standard to which the facility is registered
- Serves as a cross-reference among facility procedures
- Serves as quality system reference document for the ISO 9000 registrar, as well as other outside entities, such as customers, prospects, and investors, which the facility may designate
- Provides an adequate description of the quality system

Since many American firms already have quality manuals, most American managers are accustomed to the idea. But traditional U.S. quality manuals are strikingly different from the typical ISO 9000 version. For one thing, the typical American quality manual is huge. The ISO 9000 quality manual, on the other hand, need be 20 to 35 pages long at most. Such compression is achievable because the ISO 9000 quality manual does not recite or detail procedures. It simply makes reference to them. Another difference is purpose. Sadly, many American quality manuals are management wish lists with little resemblance to actual facility practice. The ISO 9000 quality manual is just the reverse: a one-to-one reflection of the facility's quality system as it is operated every day.

Contents of the manual

The standard suggests that methods be established for making changes to the manual and controlling its distribution. The format is left up to the individual facility. However, since the usual purpose of the manual is to document the facility's adherence to the ISO 9000 quality system standard, it follows that the manual must address each element of the ISO 9000 model that the facility adopts.

Therefore, if the facility adopts ISO 9001: 2000 and its quality system addresses all 23 of its elements, then the manual also should address the requirements of those 23 elements, specifying in general how the quality system conforms to them. The manual need not be structured in the same order as the standard, but it is often easier to adopt that structure. Whatever system is chosen, the manual must be cross-referenced to each of the 23 elements.

Although the manual may be thought of as the ultimate procedure or work instruction and, in many facilities, is actively used as a training and initiation document, it usually does not get as detailed or specific as a procedure. It is, rather, a general document, more philosophical than procedural. It expresses general principles and refers the reader to specific procedures and work instructions as needed.

In addition to the material addressing the standard's requirements, the typical ISO 9000 quality manual often includes the following sections:

- A brief statement of the facility's quality policy, in conformance with the standard's requirement that the quality policy be publicized to all facility employees, and quality objectives
- A brief facility profile outlining the organization's quality practices, which is useful when the quality manual is presented to potential customers and suppliers
- Structure of facility, including responsibilities
- Description of quality system
- Optional facility mission statement
- Distribution list for controlled circulation
- Centralized list of facility procedures

In addition to playing a vital role in developing and operating the facility's quality system, the manual is critical to the ISO 9000 registration process. The quality manual is examined in detail in Chap. 12. A sample manual appears in Appendix C.

Quality plans and operating procedures

The second and more detailed tier of quality system documentation consists of quality plans and operating procedures. The standard calls for developing documents where their absence would adversely affect quality. As a practical matter, many, if not most, facilities develop some sort of quality plans and/or operating procedures if they are not already in place.

Quality plans are defining documents, usually tied to specific product or service output. They describe

- The quality objectives for output
- Responsibility and authority for meeting those objectives
- Resources required
- Sequence of activities, cross-referencing procedures, work instructions, and related matters (see below)
- Quality monitoring methods
- Quality characteristics to be monitored
- Recordkeeping requirements

Operating procedures are still more detailed. Procedures define how activities are to be conducted, controlled, and recorded. They should include cross-references to the related quality plans and, where appropriate, to the quality manual. Operating procedures are not necessarily job descriptions, although they can be developed as such. They are specific to the process, not to an individual.

Many facilities develop quality plans and operating procedures by flowcharting the process, searching for the appropriate areas, and then creating the necessary documents, along with the associated controls. This network of documents also may include an organization chart in order to ensure that the relationship of process elements, quality system, and documentation is thoroughly understood by all people in the company.

Third-level documentation

Depending on the facility, there may be levels of quality system documentation beyond quality plans and procedures. Examples include operating instructions, manuals, and books. These levels of documentation are highly specific to individual workers and tasks and, in keeping with ISO's dictum against burdensome documentation, should be kept to a minimum.

Managing quality system documentation

As you have seen, the ISO 9000 standard requires quality system documentation but qualifies its discussion of documentation with terms such as *should* and *appropriate.* In addition, the standard shies away from making hard-and-fast rules on how to generate and use the documentation. Instead, as usual, the standard enumerates goals and guidelines and leaves many specifics up to the enlightened wisdom of facility management.

Once again, the goals have a common-sense orientation:

- Procedures and other elements of quality system documentation should be agreed to. These elements, up to and including the quality manual, are best created with the advice and counsel of the people actually doing the work.
- Documentation should be understood, i.e., written in straightforward language, legible, dated, clear, readily identifiable, and approved by authorized personnel. It also must be accessible to the employees who need it, reviewed for revision, and removed when obsolete.

Document Control, Quality Records (ISO 9001: 1994, 4.5, 4.16; ISO 9004-1, 17; ISO 9001: 2000, 5.5.6, 5.5.7; ISO 9004: 2000, 5.5.5)

Checklist of requirements:

- Documents relating to quality system requirements must be created in accordance with documented procedures.
- Personnel whose activities affect quality must have access to current editions of documents pertinent to their quality-related functions.
- The current version of each document must be easily identifiable through the use of a master list.
- Changes to such documents shall be made and approved according to procedures.

According to the standard, quality documentation is composed of drawings, specifications, blueprints, procedures, electronic media, and many other types of records. All are intended to document the quality level of output, as well as the overall performance of the quality system.

The standard outlines a few guidelines for the facility's document control scheme:

- Quality documents, as well as any subsequent revisions, are to be reviewed and approved by authorized persons.

- Documentation and records should be legible, dated, identifiable, and orderly.
- Current editions of quality documents must be available where needed, i.e., where operations affecting quality are being performed.
- Quality records and documentation should be stored in a way to prevent damage, loss, and deterioration.
- Quality documents and records should be retained as long as they are needed to facilitate analysis of quality trends and to determine the necessity for and effectiveness of corrective action. These retention periods should be documented.
- To prevent obsolete editions from being used, a procedure must ensure that outdated documents are promptly discarded or suitably identified.
- To further preclude the risk of inadvertently using obsolete editions, a master control list or some other type of list must itemize current editions of quality documents.

The ISO standard is dead serious about document control. This is evident in its repeated use of the term *shall* instead of *should*. On the whole, however, the standard presents *objectives* for the quality system without telling the facility how it must achieve them. It only requires that the facility control changes through documented procedures.

Chapter

7

Dealing with Subcontractors

Virtually every type of business acquires some form of product or service from outside suppliers. Input ranges from the most basic raw materials, to finished goods of incredible complexity, to a host of intricate service packages. In this era of decentralization, spin-offs, and specialization, vertical integration is mostly out of fashion. Businesses depend on outside suppliers more than ever. Many are, in fact, virtual hostages to them.

This locked-in dependency has a critical impact on quality. As quality gurus have counseled for many years, the quality of a process's output, product, or service begins with the quality of the process's input, the products or services supplied to it.

The quality of input has a direct and measurable impact on the scope, rigor, and effectiveness of the facility's quality system. In other words, the higher the quality of the input, the less extensive the facility's quality measures have to be. Moreover, excellent input almost by definition improves the quality of output.

Optimally, the input should be of such consistently high quality that the facility can incorporate it into its process on arrival, without any inspection or other measures. This creates economies that can be exponential, including

- Zero inspection costs
- Lower inventory costs with just-in-time delivery
- Less variability in critical characteristics, improving quality at each stage of the process

On the other hand, an ineffective supplier system, typified by multiple or redundant suppliers, emphasis on bidding wars, and adversarial relationships, puts excruciating pressure on the facility's quality system and virtually guarantees a consistently lower level quality of output.

The ISO 9000 quality system standard recognizes the vital impact suppliers or *subcontractors* have on a facility's quality system. Its requirements for subcontractor relationships are based on a philosophy of open and close communication between facility and subcontractor. These guidelines are meant to create a clear understanding of requirements on both sides. An ISO 9000 procurement system has two stated outcomes, with obvious benefits to both sides:

1. Creation of continual improvement between supplier and subcontractor
2. Avoidance or quick settlement of quality disputes

The requirements and guidelines occur under the area of purchasing, which is explored below.

Purchasing (ISO 9001: 1994, 4.6; ISO 9004-1, 9; ISO 9004-2, 6.2.4.3; ISO 9001 and 9004: 2000, 7.4)

A working relationship should be established with subcontractors to support quality improvements and to avoid or quickly settle quality disputes. Procurement of input should be planned and controlled. Procurement requirements should include

- Purchase orders, with descriptions or specifications
- A system for selecting qualified subcontractors
- Agreement on quality requirements and quality assurance requirements
- Agreement on quality assurance and verification methods
- Provision for settlement of quality disputes
- Controls for incoming product or service
- Maintenance of appropriate records for incoming product or service

Procurement control system

Procurement control requirements are quite similar to the quality system requirements outlined elsewhere in the standard. They follow the ISO 9000 principles of planning, control, and documentation.

The first and most important requirement is that the facility establish a close working relationship and feedback system with each subcontractor. Within this overall requirement exists an array of others.

1. *Subcontractors must be chosen for their ability to meet quality and product or service requirements.* Before ordering from a subcontractor, it is only good common sense to make sure that the subcontractor has the ability to meet the facility's needs. The standard requires the facility to use a system for verifying subcontractor capability. This system may include

- On-site evaluation of the subcontractor's capability and/or quality system
- Prior history with the same or similar input
- Evaluation and testing of sample input
- Published experience of other users

The standard also suggests that the facility require subcontractor registration to one of the ISO models, ISO 9001, 9002, or 9003, as a strategy for meeting this requirement.

2. *Requirements should be clearly defined and understood.* The facility and subcontractor should define and agree on the crucial characteristics of the input to be provided. These characteristics are usually set forth in one or more standard documents, such as purchase orders, accompanied by blueprints, drawings, specification sheets, and other supporting documents. It is up to the facility to ensure that all vital details are communicated to *and understood by* the subcontractor.

Therefore, the standard suggests that the facility employ procedures governing measures necessary to attain the required level of understanding. These procedures may involve creation and verification of purchase orders and supporting documents, arranging and managing facility-subcontractor meetings, and other measures, as appropriate.

3. *Facility and subcontractor should agree on quality assurance that the subcontractor is to provide.* An array of options is available here:

- None. The facility conducts 100 percent inspection of input.
- Subcontractor conducts inspection and testing, either 100 percent or through lot sampling, providing test results to the facility.
- Subcontractor provides various process control records, including inspection and test data, with shipments.
- Facility relies on subcontractor quality assurance without verification.

4. *Facility and subcontractor should agree on methods of verifying conformance.* The facility should ensure that the subcontractor clearly understands how the facility will verify that input conforms to requirements. Such understanding will smooth communication and reduce disputes. To fulfill this requirement, the facility may have to implement formal documented procedures. The facility and subcontractor may wish to exchange technical data with subcontractors and engage in joint training in various statistical techniques and improvement methods.

5. *Facility and subcontractor should agree on systems and procedures for dealing with routine and nonroutine matters, especially disputes affecting quality.* This is a simple matter of arranging for open channels of communication between the facility and the subcontractor.

Controlling input

The ISO 9000 standard is very specific about the facility's obligation to control

- Materials of any nature that are known to be nonconforming to requirements
- Materials whose conformance status is unknown

In essence, the standard requires the facility to make certain, by means of documented procedures, that such materials remain separate from conforming materials.

This general requirement applies to input in general. The standard requires that the procedure for receiving input be planned carefully. As such, the procedure should ensure that received input is controlled and kept separate until its conformance status has been determined.

If input is accepted by means of subcontractor verification or other methods prior to receipt, then this requirement may be moot. Many facilities, however, employ a receiving inspection procedure. Those which do are admonished to balance the costs of inspection against the consequences of inadequate inspection when developing their inspection system. The standard also requires such facilities to

- Select inspection characteristics carefully
- Ensure that inspections are carried out by trained and qualified personnel using inspection instruments that are properly calibrated

Quality records

As with most other aspects of the quality system standard, facilities are obligated to maintain quality records of subcontractor activities. The standard specifically mentions records containing historical data that aid in assessing supplier performance and evaluating quality trends.

Chapter

8

Dealing with Customers

American managers who are implementing ISO 9000 quality systems often make the same beleaguered comment: "ISO 9000 doesn't address customer satisfaction."

It is easy to see how some would reach this erroneous conclusion. The old contractual models, ISO 9001, 9002, and 9003, have no elements dedicated to customers per se. ISO 9001: 2000, on the other hand, contains Elements 5.2, Customer Focus, 7.2, Customer-Related Processes, and 8.2.1, Customer Satisfaction. Element 6.1, Provision of Resources, states that the organization must provide and determine the resources needed to address customer satisfaction.

ISO 9004-1 and 9004-2, the old guidelines documents for manufacturing and service applications, devote extensive attention to the role of customers in the facility's quality system, as does ISO 9004: 2000, which includes customers among interested parties in Elements 5.2, 7.2, and 8.2.1.2. All versions of the quality loop include customers as an element (see Figs. 5.1, 5.2, and 5.3). In fact, all versions begin and end with customers.

These schematics attest to the importance ISO 9000 places on customers in developing and managing of the facility's quality system. The requirements and guidelines are presented among a number of topic headings, all explored in detail below:

Quality in Marketing

Customer Communication

Contract Review

Control of Customer-Supplied Product

Servicing

Quality in Marketing (ISO 9004-1, 7; ISO 9004-2, 6.1; ISO 9001 and 9004: 2000, 5.2, 7.2)

Checklist of requirements:

- Marketing establishes defined and documented quality requirements for output.
- Marketing creates product or service brief.
- Marketing operates an information monitoring and feedback system to evaluate and communicate all customer requirements.
- Marketing ensures that all relevant facility functions agree that they have the capability to meet customer requirements.

In most organizations, marketing is the major means of communication between the facility and the customer. Recognizing this, ISO 9000 places on marketing the responsibility for

- Identifying customer requirements
- Translating these requirements into initial product or service specifications
- Translating product or service features into advertising
- Gathering, analyzing, and communicating customer feedback

Where appropriate, facilities should have systems and procedures for carrying out these activities.

Identifying customer requirements

According to the ISO 9001 standard, marketing is responsible for identifying customer requirements, including, of course, all requirements with respect to quality. However, the responsibility goes well beyond this. Marketing also must take the lead in the following areas:

- Determining the need for products or services
- Identifying relevant legislation, standards, and codes
- Defining the market for products or services
- Determining and evaluating competitive offerings
- Defining the details of customer needs and expectations, including implicit and/or unstated needs and expectations
- Identifying specific quality verification and assurance requirements and characteristics

These activities should be conducted not just in the initial development phase but also on an ongoing basis for existing products or services.

Creating the product brief

Marketing should, by procedure, create a *product brief* that formally summarizes its findings with respect to product or service features, especially as they relate to quality. The product brief is, in effect, a preliminary set of specifications that marketing communicates to the other areas of the facility. It should specifically relate these specifications to the facility's internal capabilities.

The product brief is used as the basis for designing the facility's product or service (see Chap. 9). It leads to the development of formal and more detailed specifications and quality characteristics. These characteristics, in turn, act as a baseline or benchmark against which the resulting output can be measured.

The specifics of the product brief can, of course, vary tremendously from facility to facility. The standard makes no specific requirements. It only includes areas for consideration, such as

- Performance and sensory characteristics
- Statutory or regulatory requirements
- Installation and use issues
- Packaging
- Quality verification and assurance factors

Advertising

ISO 9004-2, the guidelines document for service functions, makes specific mention of quality in advertising. It calls for a facility to make sure that its advertising

- Reflects the product or service specifications
- Takes into account customers' perceptions of quality
- Refrains from offering exaggerated or unsubstantiated claims

These guidelines only appear particularly in ISO 9004-2. ISO 9001: 2000, Element 7.2, requires organizations to identify and implement arrangements to communicate with customers relating to product information. ISO 9004: 2000 offers no advertising guidelines.

Customer feedback

Finally, marketing is responsible for closing the customer loop by operating a continuous information monitoring and feedback system. The system should operate in accordance with defined procedures, focusing specifically on feedback relating to the quality of output. This information should be communicated to appropriate departments of the facility.

The information should be used for the following purposes:

- Determining customers' quality experience, as compared with quality expectations
- Identifying potential output improvements
- Serving as part of the continuous evaluation of customer requirements called for earlier

Customer Communication (ISO 9004-2, 5.5; ISO 9001: 2000, 7.2.3; ISO 9004: 2000, 7.2)

Checklist of requirements:

- Management should establish and regularly review methods of interaction between the facility and the customer.

The customer interface section of ISO 9004-2, as well as ISO 9001: 2000, Element 7.2.3, Customer Communication, expands on some of the themes of the marketing guidelines described earlier. While these provisions do not appear in the other parts of the old standard, they are well worth incorporating, as appropriate, into every quality system, owing to the addition of the new customer communication element (7.2.3) to ISO 9001: 2000.

Communication between a facility and its customers is important, particularly when it comes to providing product information, handling contracts and orders, and determining customer perceptions of quality.

The facility should recognize this by

- Instituting procedures to ensure that interactions with customers are as effective as possible
- Routinely reviewing methods of promoting customer contact

Difficulties in communication should be given prompt attention, especially when they involve such issues as

- Product and/or service information, features, benefits, availability, timeliness of delivery, and cost
- Enquiries, contracts, or order handling, including amendments
- Customer feedback, including complaints, service, and warranty issues
- Customer needs and expectations

Contract Review (ISO 9001: 1994, 4.3; ISO 9001: 2000, 7.2.2; ISO 9004: 2000, 7.2)

Checklist of requirements:

- Facility shall maintain procedures for conducting and documenting contract reviews.

This element of ISO 9001: 1994 has strong parallels to the quality in marketing requirements discussed earlier, with both appearing in the same element of ISO 9001: 2000. Since these requirements pertain to contractual relationships, they are restated much more pithily in the context of a contract between facility and customer.

The contract review element, known as review of product requirements in ISO 9001: 2000, obligates the facility to maintain procedures to evaluate customer requirements and compare them to the facility's capabilities. The reviews must ensure that

- Requirements are adequately defined and documented.
- Requirements are within facility capabilities.
- Discrepancies are resolved.
- Contract amendments are incorporated into the appropriate functions.

As might be expected, this section also requires that the facility maintain records of these contract or product requirement reviews.

Control of Customer-Supplied Product (ISO 9001 4.7; ISO 9001 and 9004: 2000, 7.5.3)

Checklist of requirements:

- Facility shall maintain documented procedures to control and protect materials provided to it by customers.

This section of ISO 9001 applies to a facility that, as part of its process, receives materials, products, services, or other property owned by customers for inclusion in the output, which is then supplied to them. For example, a book manufacturer (supplier) is in the business of printing and binding books for sale to a book publisher (customer). As part of the process, the publisher supplies the book manufacturer with a computer disk containing the text that is to be included in the final output. That disk is *customer-supplied product* (ISO 9001: 1994) or *customer property* (ISO 9001: 2000).

Facilities that receive such customer-supplied product or other property must monitor and secure it at each point during which it is under the facility's control. However, verification by the supplier does not absolve the customer of the responsibility to provide acceptable product.

The system maintained by the facility should

- Document suitability of the product or other property for the intended purpose, via inspection or other means.
- Keep customer-supplied product or other property safe and secure.
- Report to the customer any occurrences of nonconformance, shrinkage, damage, loss, or other problems affecting customer-supplied product or other property.

Servicing (ISO 9001: 1994, 4.19; ISO 9001: 2000, 7.5.1; ISO 9004: 2000, 7.5.6)

Checklist of requirements:

- Facility shall maintain procedures to ensure that contractually required servicing is performed according to requirements.

At 26 words, servicing is the shortest element in ISO 9001: 1994 and gets even less attention in ISO 9001: 2000. It usually applies only to facilities whose customer contracts oblige them, by warranty or other means, to provide postsale servicing.

This element obligates the supplier to establish and maintain procedures for performing all contractual service and verifying and reporting that the service meets specified requirements. Servicing requirements may include customer-use documentation, installation instructions, postsale support facilities with staff and equipment, and field measuring and test equipment.

When postsale servicing is part of a customer contract, the facility's product design function must anticipate service requirements with appropriate procedures, mechanisms, personnel, and skills and ensure that these are in place well in advance of initial deliveries. An early warning system should be in place to ensure timely detection of service needs.

Servicing is addressed in ISO 9001: 2000, Element 7.5.1, which requires an organization to control production and service operations through implementing defined processes for release, delivery, and applicable postdelivery activities.

Chapter

9

Designing and Producing the Output

So far we have looked at what ISO 9000 says about the components of the quality system, its documentation, and relations with subcontractors and customers. These elements of the quality system standard are more general and external. They do not directly address the output (product or service) being produced.

Still, these elements are critical to the success of other, output-specific provisions of the standard. Ultimately, they are quite critical to the quality of the output itself. They are what makes ISO 9000 a quality *system* standard, embracing every aspect of the facility that affects the quality of output, as opposed to a quality tool or technique, which focuses on quality of the output only.

ISO has plenty to say about output, though. Now, having covered these broader issues, we can narrow the focus a bit. This chapter and the two that follow explore the quality system elements required for the creation of output: designing it, producing it, controlling it, and ensuring its quality.

Under the old standard, facilities that design and produce output, and which intend to become registered to ISO 9000, adhere to the 20 sections of ISO 9001 and the associated guidelines in ISO 9004-1. Facilities that produce output designed by others adhere to 19 of those sections, as contained in ISO 9002. Under the 2000 revisions, all facilities will register to ISO 9001, which can be tailored to delete inapplicable requirements under Element 1.2, Permissible Exclusions.

The ISO 9000 standard recognizes the critical impact of the design function on the quality of output. Its requirements spell out an integrated, documented, and controlled design process. This process translates the customer requirements listed in the marketing brief (see Chap. 8) into technical specifications and designs that ensure output that meets those requirements. The ISO 9000 design process

- Stays in touch with customer requirements
- Takes into account the capability of the process
- Incorporates identifiable and measurable quality characteristics
- Specifies acceptance criteria and measurement methods for those quality characteristics
- Builds in quality control mechanisms
- Remains adaptable to controlled and disciplined change

The ISO 9000 quality system dovetails this design process to a production process that is planned, controlled, and documented at every juncture. Adherence to the standard results in a design and production process that

- Creates output consistent with customer expectations
- Takes pertinent safety and product liability issues into account
- Generates a predictable return on investment
- Includes systems for managing internally or externally driven change

The various parts of ISO 9000 describe quality system requirements for design and production in several different elements, as explored in detail below:

Design control

Process control

Product safety

Design Control (ISO 9001: 1994, 4.4; ISO 9004-1, 8; ISO 9004-2, 6.2; ISO 9001 and 9004: 2000, 7.3)

Checklist of requirements:

- Management should establish a design program appropriate to the application, complexity, and innovation of the output.
- Management must ensure that the design function is aware of quality responsibilities and goals.

- Design plans that describe design activities and assign responsibilities must be prepared. The plans must be updated throughout the design process.
- The design process should create technical specifications that result in output that meets customer requirements, safety and environment regulations, and facility return-on-investment objectives.
- Designs should unambiguously specify quality characteristics, measurement and test methods, and acceptance criteria.
- The design process should include objective and periodic design reevaluations and systematic and critical design reviews at prescribed intervals.
- The design process should end with a readiness review to verify that all elements deemed necessary for quality production and distribution are in place.
- The quality system should include documented procedures for processing, in a controlled manner, routine and emergency design changes throughout the life cycle of the output.
- Consideration should be given to applicable regulatory and legal requirements.

The design program

The standard includes a basic and quite obvious requirement that the design program be time phased, meaning planned and disciplined, and consistent with the application, complexity, and innovation of the output. In other words, the newer, riskier, less standard and less derivative the output concept, the more planned, deliberate, and reflective must be the design process.

Design of services should define the service itself, the delivery of the service, and the quality procedures for controlling service characteristics.

Design personnel

Management is responsible for delegating design responsibility to specific people or entities. This requirement may seem obvious, but it is the standard's way of reinforcing the concept that the design process should be planned and disciplined.

The standard also requires management to keep the design function aware of its responsibilities for quality. This reiteration of the management responsibility element serves to underline the standard's position that the design process has a powerful and definitive effect on quality. In fact, ISO 9004-2 clearly states at Element 6.2.2 that "*prevention* of service defects at this stage is less costly than *correction* during service delivery."

Technical specifications

According to the standard, the chief goal of the design process is to translate customer requirements into technical specifications for the output. These specifications should provide clear and definitive instructions for

- Quality aspects of the design
- Characteristics important to quality
- Data for procurement from subcontractors
- Instructions for performance of work
- Guidelines for verifying conformance to requirements
- Contingency plans

In all of the preceding, the design function is cautioned to take into account any pertinent legal, regulatory, contractual, and environmental considerations.

Output testing and measurement

Just as designs must specify quality aspects and characteristics, they also must specify how the facility will monitor the results of its efforts to maximize the quality of these characteristics. The standard states that measurement methods and acceptance criteria for all quality characteristics should be clearly specified during both design and production phases.

The standard recommends that design ensure the monitoring of quality characteristics by including, for each

- Tolerances, attributes, and target values
- Measurement and test methods, equipment, and computer software
- Acceptance and rejection criteria

Design qualification, validation, verification, and review

To ensure that the design process is a rigorously disciplined one and results in designs that conform to customer requirements, the ISO 9000 standard requires regular and documented design review activities. One of these is a *design qualification and validation* process. This process, to be conducted at significant stages of the design and development cycle, subjects prototypes, samples, and related materials to appropriate tests. These tests may study such issues as:

- Performance, durability, and safety
- Conformance of design features to design goals
- Confirming verification of original calculations

A related activity is *design verification,* which ensures that design output meets the requirements of design input. It is required at appropriate design stages. Verification measures, which must be recorded, may include

- Comparison with proven designs
- Tests, demonstrations, and alternative calculations
- Review of design stage documents before release

A broader type of analysis is the *design review.* These reviews are, according to the standard, formal, documented, systematic, and critical examinations of design results. They consider, in detail, three major design elements:

- Factors affecting customer needs and satisfaction
- Factors affecting output specification and service requirements
- Factors affecting process specifications and service requirements

The purpose of design reviews is to spot problem areas and initiate corrective actions. The principal objective is, once again, ensuring that the final design meets customer requirements.

The standard requires regular reviews to be conducted after each significant phase of design development. The final review, at the conclusion of the design process, is documented, as appropriate, with final specifications, drawings, or in the case of a service, delivery procedures.

This final design serves as a baseline against which resulting output is compared. It should undergo review and approval by appropriate management levels. Finally, it should be accompanied by a *market readiness review* to verify that required production capability, field resources, and other support systems are in place to handle the output.

Design changes

Design changes are a fact of life. Inefficient means for handling them are a major cause of bloated development costs, competitive shortfalls, extended cycle times, and ultimately, loss of market share. Deficient design change procedures also can have a devastating effect on both quality and the facility's efforts to meet customer requirements.

Under ISO 9000, the facility must have a procedure to control the release, change, review, and use of documents that define the *product baseline,* also known as the *final design* and/or *service specification.* Design changes amount to changes to the product baseline. The standard calls this process *configuration management* and intends for it to be planned and disciplined as well.

The procedures should provide for

- Appropriate approvals
- Scheduled activities for planning and implementing changes
- Updating of documents, work instructions and other materials, and removal of obsolete documents
- Follow-up to ensure that all specified activities have been carried out

Configuration management procedures are expected to be adequate to handle emergency changes needed to ensure that output conforms to customer requirements.

Design review

The standard recognizes that things change, a basic fact of life. Today's ideal design, even one developed under an ISO 9000 quality system, may be grossly inferior tomorrow. Certainly, competitive pressures and changing markets will result in changing customer expectations. Designs must change, too, if a facility is to continue to meet the primary quality goal of satisfying customer expectations.

Thus ISO 9000 calls for facilities to conduct regular *design and development reviews* to ensure that the design still meets customer requirements. The reviews should take into account

- Field experience
- Surveys, research, and other customer feedback
- Competitive analysis
- New technologies

The findings of design and development reviews can result in routine or emergency design changes, process changes, or new output development initiatives.

Documentation

Finally, the standard requires that all design verification and review activities be given appropriate documentation.

Process Control (ISO 9001: 1994, 4.9; ISO 9004-1, 10, 11; ISO 9004-2, 6.3; ISO 9001: 2000, 7.1, 7.5.1, 7.5.5; ISO 9004: 2000, 7.1, 7.5.1)

Checklist of requirements:

- Production capability should be monitored, especially those aspects affecting quality, to ensure that production can meet quality requirements.
- Input materials, input processes, and production equipment should be maintained, controlled, and verified, as appropriate, to ensure consistent conformance to requirements.
- Production, especially aspects affecting quality, should be carried out under planned, controlled, and documented conditions.
- Activities to verify quality status should be conducted at critical intervals on a planned and documented basis.
- Special controls and procedures should be implemented for processes that are highly critical to quality and/or difficult to monitor effectively.
- Procedures for making process changes should be thoroughly planned, documented, monitored, and evaluated.
- All quality documentation pertaining to production should be maintained in accordance with documented procedures.

Production capability

It may seem obvious, but each production process has a limited level of capability. A process forced to perform beyond its capability is not a quality process creating quality output. It is, in fact, a sign of management ignorance.

Under ISO 9000, the capability of a process to produce output to specification is defined and understood at all times. Management is required to implement procedures defining general capability. The design control elements (ISO 9001: 1994, 4.4 and ISO 9001: 2000, 7.3) address this issue, and it is also examined here for quality in production.

In addition to general capability, the standard obligates management to identify process operations specifically concerned with output quality. The capability of these operations must be monitored closely to ensure that quality goals are met. The standard also obligates management to encourage efforts to develop new methods for improving quality and process capability.

Production materials and equipment

The subcontractor sections of the ISO 9000 standard address the control of supplied input. Here, under production materials and equipment, the standard mentions auxiliary input items, such as utilities, and the environment. The standard calls for such factors to be controlled and verified if they are shown to affect quality.

The same principles and requirements apply to materials and equipment used in production. The standard requires that these be

- Verified as conforming to specifications and quality standards before being introduced into the process
- Maintained and used under planned and controlled conditions, with traceability if needed
- Subject to regular reverification procedures, including appropriate preventive maintenance

Production control

As with virtually everything else in an ISO 9000 quality system, production is expected to be planned, controlled, and documented.

1. Planning ensures the appropriate manner and sequence of production.
2. Control covers all aspects of production, including materials, equipment, procedures, personnel, supplies, and environment.
3. Documentation is essential, but only to the necessary extent.

Specifically, the standard calls for the use of documented work instructions. These should, where appropriate, govern production activities that affect quality. Common work instructions, covering identical activities throughout the facility, can be used. Specialized work instructions, applying to specific process steps, also can be implemented.

Work instructions should

- Describe criteria for determining satisfactory work
- Define work standards
- Be controlled in accordance with document control procedures (see Chap. 6)

Verifying quality

To minimize errors and maximize production, verification of quality status, of both output and process elements affecting its quality, should be ascertained at important points in the production sequence. The verification requirement specifically refutes a common criticism that ISO 9000 is inspection-based. ISO 9000 does not require or even prefer inspection. It requires verification of quality status via process control methods, including control charts and statistical sampling procedures and plans. If such controls are not feasible or practical, then product control methods, such as inspection, must be employed.

The standard discourages petty or frivolous verification systems by stating that verifications should relate directly to either finished output specifications or some internal process requirement. Verification requirements are examined in further detail Chap. 10.

Special processes

The standard calls for special measures to be employed to manage quality in certain special process areas. These are processes in which

- Output characteristics cannot be measured easily or economically.
- Results cannot be fully verified by inspection or testing.
- Defects can be detected only after output is in use.

The implicit requirement is that management take steps to identify these processes and their quality issues. Then, to manage quality in these special processes, the standard suggests intensified quality system measures, including

- Measurement instruments that provide higher accuracy and less variability
- Higher employee qualification standards, training regimens, and possibly certification requirements
- Specialized physical environments to increase conduciveness to quality

Process changes

Just as designs undergo change, so do processes. And again like design changes, process changes can be chaotic or smooth. As you might expect, ISO 9000 comes down firmly on the side of smooth. Accordingly, the standard calls for process changes to be controlled as follows:

1. Responsibility should be clearly designated by management.
2. The customer should be consulted, if appropriate.
3. Implementation should be conducted in accordance with defined procedures.
4. Changes should be documented.

Further, under ISO 9000, subsequent output must be evaluated to ensure that process changes have not adversely affected critical characteristics.

Product Safety (ISO 9004-1, 19)

Checklist of requirements: The facility should

- Identify all safety aspects of the quality of output.
- Carry out design safety testing and document test results.
- Incorporate applicable safety standards into output design, test output for effectiveness of safety measures, and test all user communications for clarity and accuracy.
- Provide instructions and alert users to any warnings regarding known hazards.
- Institute output traceability to ensure effective recall if safety subsequently is found to have been compromised.
- Consider creating an emergency plan in the event of a product recall.

Safety and product liability are crucial and volatile issues, especially in today's litigious business environment. Every facility wants to maximize the safety of its customers, obey safety regulations, and minimize exposure to product liability actions.

Where safety and product liability are factors, the ISO 9000 quality system standard treats them as a special category of critical quality characteristics. The ISO 9000: 2000 standards, on the other hand, do not address product safety issues directly. In ISO 9004-1, facilities for which safety is an issue and/or those which are subject to safety regulations of any type are expected to

- Know the safety issues and regulations
- Address them with appropriate specifications throughout the design process
- Test for them as part of ongoing verification activities
- Document all safety actions

Facilities are also advised to institute traceability measures in order to recall any output that, subsequent to delivery, is suspected of nonconformance to safety standards. Traceability is examined in more detail in Chap. 11.

Chapter

10

Ensuring the Quality of Output

Now we reach the core of the ISO 9000 quality system. The requirements covered in earlier chapters set the stage. Here we will examine what the standard says about ensuring the quality of output as it moves through the process.

There is nothing revolutionary in these technical requirements. Like the rest of the ISO 9000 standard, the guidelines in this area are mainly good common sense. The approach differs somewhat between the service and the nonservice parts of the old standard, but the essential strategy is the same:

1. Know the critical quality characteristics and acceptance standards.
2. Verify conformance to standards at critical process points.
3. Use procedures to ensure the adequacy and accuracy of measuring equipment.
4. Maintain appropriate identification of verification status of output at all points.
5. Use appropriate statistical techniques.
6. Identify output appropriately.
7. Segregate nonconforming output.
8. Determine causes of nonconformances.
9. Devise, implement, and verify corrective actions.

The last four provisions are examined in Chap. 11. The provisions we will look at here are worded more specifically, but not much more, than other parts of the standard. However, they remain deftly phrased and generally applicable. It is the genius of the ISO 9000 standard that it can be adapted and applied to such a broad range of products and services. The tailoring option of ISO 9001: 2000 will make it even more flexible.

Notice that the strategy just described is a cyclic prescription for *continual improvement.* ISO 9000 is not a system for sorting acceptable output from defective output. It is a system to ensure (1) that customers receive output that meets their criteria and (2) that process elements that do not contribute to that goal are constantly improved.

The process of creating services is much different from the process of creating products. As a result, ISO 9004-1 (products) and ISO 9004-2 (services) differ rather significantly. ISO 9004: 2000, on the other hand, is more generic, applying to both products and services.

The first section of this chapter focuses on the service quality guidelines discussed in ISO 9004-2. The remaining sections will take up the product quality requirements outlined in both ISO 9004-1 and ISO 9001.

Even if your facility is strictly involved with products, you should review the ISO 9004-2 requirements because they are much more aggressive regarding corrective action and continuous improvement. In real life, almost no facility is so purely a producer of products that it can ignore quality service successfully. In fact, it can be argued that a product company cannot succeed on the quality of its output alone. Without quality service to go with it, the company is doomed to fail.

ISO 9000 describes quality system requirements for ensuring the quality of output in several different sections, as explored below:

Quality in the production of services

Inspection and testing

Control of inspection, measuring, and test equipment

Inspection and test status

Statistical techniques

Quality in the Production of Services (ISO 9004-2, 6.3, 6.4)

Checklist of requirements:

- Services should be delivered according to a documented specification.
- The service delivery process should undergo a continuous assessment at critical process points using documented and verified measurement and testing methods.
- The service delivery process should aggressively collect and analyze customer assessments of service quality.

- Nonconformities should be recorded and analyzed and then subjected to appropriate, documented corrective action.
- Management should operate a system to gather data on service performance and analyze it for deficiencies, corrective actions, and opportunities for improvement.
- Management should operate a program to continuously improve service quality.

Service specification

This requirement echoes the management responsibility guidelines discussed earlier. Management is expected to define and document the service delivery process. Appropriate personnel should be assigned specific responsibilities, including delivering the process, assessing its effectiveness, obtaining feedback, and performing corrective actions.

Internal assessment of service quality

The service delivery process should include quality control procedures at critical process points. These procedures should include

- Monitoring key process activities, particularly by employees delivering the service, to detect deficiencies and ensure customer satisfaction
- Definitive quality assessment by facility personnel dealing with customers, for later comparison with customer quality assessment
- Final supplier assessment to ensure that the supplier's perspective of quality service was delivered

Throughout, the system should employ documented and verified measurement and testing methods.

Customer assessment of service quality

The standard calls customer assessment the ultimate measure of service quality. However, objective customer assessment is not easy to obtain. Passive methods, such as depending on customer complaints, are not reliable. Customers can be dissatisfied to the point of taking their business elsewhere, without bothering to complain. Effective corrective action is impossible without ongoing and objective gathering of customer quality assessments.

Thus the standard calls on facilities to operate an ongoing system to assess and measure customer satisfaction. The system should study not only how well the service delivery meets the service procedure requirements but also how well the documented service procedure or *service brief* meets prevailing customer needs. A facility may carry out its documented service delivery procedures faithfully and believe, therefore, that it is delivering flawless quality. Unfortunately, the customer will not see it this way if the service delivery process does not fully address customer needs.

Elements of this system include

- Gathering both positive and negative feedback
- Assessing the likelihood of future business
- Comparing customer assessments to the facility's own assessment

Corrective action

This is the standard's vehicle for continual improvement. It is specified in the product-oriented ISO 9004-1 but is explained more thoroughly in ISO 9004-2. In ISO 9001 and 9004: 2000, it is covered by Element 8.5.2.

ISO 9004-2 sets another management responsibility guideline, under which the quality system should define responsibility and authority for corrective action. Furthermore, everyone in the organization is responsible for

- Identifying nonconforming services
- Reporting nonconforming services
- Doing so before customers are affected, if at all possible

Once nonconformances are recorded and analyzed, corrective action should be taken. The standard describes two levels of corrective and preventive action:

1. Immediate action should be taken to satisfy customer needs.
2. Longer-term action is called for to identify causes, formulate and implement corrections, and monitor for effectiveness.

Finally, information regarding actions taken must be delivered to management for review.

Service performance analysis and improvement

At this point, the standard goes beyond process-oriented quality control and corrective action. It sets forth guidelines for ensuring continuous improvement of service delivery processes. The basis for this strategy is an effective data collection and analysis activity. This should be undertaken (no surprise) in accordance with a systematic documented procedure and performed by personnel definitively assigned to the tasks by management. Data should be gathered from all relevant sources, including

- Internal assessments (quality control)
- External evaluations (all forms of customer feedback)
- External and internal quality audits

Collected data should be used to uncover systemic problems, identify causes, and create effective correction and prevention strategies. The standard calls for the use of modern statistical methods in these activities.

ISO 9000 also calls for the quality system to take aggressive and outward-looking actions to improve quality. These include

- Identifying service characteristics whose improvement would most benefit the customer
- Detecting changing market trends
- Finding ways to reduce costs while maintaining or improving service quality

Throughout, management is encouraged to seek input from all people in the organization and to recognize and reward their participation.

Inspection and Testing (ISO 9001: 1994, 4.10; ISO 9004-1, 12; ISO 9001 and 9004: 2000, 7.1, 7.5.1, 8.1, 8.2.4)

Checklist of requirements:

- The quality system must verify conformance to requirements at critical process points.
- Verification methods must be documented and the results recorded in accordance with procedures.
- Output exempted from verification should be made traceable in case of recall.
- Records must indicate whether or not a product has passed inspection and must identify the person(s) who inspected and released the product.

ISO 9001 requires documented procedures to describe inspection and testing activities. These activities are intended to verify that products are being made according to specified requirements. The inspection and testing required, and records subsequently established by these activities, must be documented in the quality plan or work procedures.

Incoming products must be held back from processing until they are verified. This verification may include inspection and testing, or it may involve recorded evidence of conformance from an acceptable subcontractor. The only exception to this verification requirement is an emergency situation, in which products are released under positive-recall procedures and verification is waived.

ISO 9000 recognizes that there are times when verification procedures must be waived. In these cases, the facility is obligated to ensure that such items are identified and made traceable in the event of a recall.

Some form of completed product verification must be conducted to ensure that output conforms to quality requirements. Where appropriate, these verifications may include references to purchase orders to confirm that specific customer requirements are met.

Whatever procedures the facility uses to verify conformance to requirements, whether inspection and testing or other means, the standard requires that they be spelled out in a quality plan or other documentation. These documented procedures must be carried out consistently, and appropriate records must be maintained.

Control of Inspection, Measuring, and Test Equipment (ISO 9001: 1994, 4.11; ISO 9004-1, 13; ISO 9001 and 9004: 2000, 7.6)

Checklist of requirements:

- Devices employed to assess conformance to requirements should be selected and controlled in order to preserve confidence in the integrity of measurements.
- Devices must be calibrated in accordance with documented procedures.
- Output assessed by devices that are found to be nonconforming should be controlled and reassessed in accordance with documented procedures.
- Records of control of calibration, maintenance of inspection, measuring, and test equipment must be maintained.

In order to ensure that products are properly measured and analyzed, ISO 9001 requires documented procedures to control the calibration and maintenance of inspection, measuring, and test equipment and the means used to perform measurements and analyses. This includes software and hardware used to verify acceptability. Suppliers may decide the frequency and extent of testing.

Controls used by suppliers must be sufficient to ensure confidence in decisions made on the basis of measurements. If a contract requires technical data from measurements, customers must be allowed to verify the adequacy of test equipment.

Devices used to obtain measurements must be appropriate to the task, used by people with appropriate training, and maintained in accordance with requirements. Such devices may include

- Gauges, instruments, and sensors
- Special test equipment
- Related computer software
- Jigs, fixtures, and process instrumentation

The standard sets forth the following guidelines concerning such devices:

- Equipment must be appropriate with respect to specified requirements, capable of required accuracy and precision in the environment in which it will be used, and handled in a way that preserves its functional integrity. Margin for error must be identified.
- Equipment and facilities associated with inspection and test operations must be protected from adjustments that alter calibration.
- Equipment must be calibrated at scheduled intervals, or before use, in accordance with documented procedures. Calibration must ensure that equipment is consistent with a nationally recognized standard or, in the absence of same, with some other documented basis for comparison. Calibration status must be identified by an indicator and documented in records.

At this point, the standard also imposes a corrective action obligation. When a device is found to be out of calibration, the facility must follow a procedure to determine the cause of the problem and take remedial action. The facility also must reevaluate previous work affected by the device for conformance to requirements.

Inspection and Test Status (ISO 9001: 1994, 4.12; ISO 9004-1, 11.7; ISO 9001 and 9004: 2000, 7.5.1)

Checklist of requirements:

- Throughout the process, conformance status of output should be identifiable via some documented means.

- The conformance status method should provide traceability to verification activity and should be defined in the quality plan or work procedures.
- Verification authority should be documented.

One overriding quality goal of ISO 9000 is to ensure that the only output reaching customers is output that conforms to requirements. Accordingly, the facility must be able to distinguish between conforming and nonconforming product not only after the final verification point but also at each critical checkpoint in the process, including installation and servicing.

As the determination is made regarding which output conforms, hopefully the vast majority, and which does not, the standard requires that the outcome of testing and inspection, conforming versus nonconforming status, be clearly identified. Such identification can be achieved through physical marking, hard copy or software records, physical location, or other suitable means.

The standard also suggests that the system for determining inspection and test status should provide traceability to the verification point where the distinction was made. Records must be kept of the authority responsible for releasing conforming output.

Statistical Techniques (ISO 9001: 1994, 4.20; ISO 9004-1, 20; ISO 9001 and 9004: 2000, 8.1, 8.2.3, 8.2.4, 8.4)

Checklist of requirements:

- The quality system should include procedures for identifying statistical techniques used in assessing process capability and output characteristics.

ISO 9004-1, Element 20.1, states that statistical methods are important in all stages of the quality loop, not just the production or inspection phases. ISO 9004: 2000, Element 8.4, states that information and data from all parts of the organization should be integrated and analyzed to evaluate the organization's overall performance. These elements in the old and new versions of ISO 9001 attest to the importance of statistical techniques in improving quality, productivity, efficiency, communications, and planning.

The standard obligates the facility to have procedures to identify statistical techniques for verifying the acceptability of process capability and product characteristics. These procedures should be employed in all areas where their use will demonstrably improve quality, such as process control, process capability, inspection planning, and defect analysis. In addition, the standard suggests that statistical techniques should be considered for nonproduction areas. Candidates here include market analysis, design, and performance assessment.

The standard does not mandate any particular statistical techniques. Suggestions for consideration include design of experiments/factorial analysis, variance/regression analysis, safety evaluation/risk analysis, cumulative sum (CUSUM) charting techniques, and statistical sampling inspection. Facilities are free to investigate and implement other statistical techniques that are shown to contribute to quality objectives.

ISO 9001: 2000 expands this element. It requires organizations to establish a system-level procedure to analyze applicable data to determine the effectiveness of the quality management system and identify where improvements can be made. Applicable data must be analyzed to provide information on

- Customer satisfaction and/or dissatisfaction
- Conformance to customer requirements
- Characteristics of process, product, and their trends
- Suppliers

Chapter

11

Controlling and Improving Output

This final phase in our examination of technical requirements closes the loop in the ISO 9000 quality system. Here we will explore the standard's guidelines for

- Using appropriate process controls to prevent nonconforming output from reaching customers
- Conveying output to customers with appropriate safeguards
- Channeling information on nonconformances into corrective actions and process improvement

Obviously, not all the requirements in these sections apply to all facilities. A maker of toothpicks, for example, may not need to implement in-depth product identification and traceability measures.

At some level, however, the other provisions apply across the board. After all, the two most critical purposes of a quality system are to

- Ensure that only verified conforming output reaches customers.
- Transform causes of nonconformances into corrective actions that result in continuous improvement.

These are the purposes specifically addressed by the ISO 9000 quality system requirements examined in this chapter. The elements for controlling and improving output, as discussed in detail below, are

Product identification and traceability

Control of nonconforming product

Handling, storage, packaging, preservation, and delivery

Corrective and preventive actions

Product Identification and Traceability (ISO 9001: 1994, 4.8; ISO 9001 and 9004: 2000, 7.5.2)

Checklist of requirements:

- To the extent deemed appropriate to meet legal, regulatory, or quality requirements, the facility should operate a documented system for identifying and tracing output from receipt and throughout all stages of production, delivery, and installation.

This section refers to three general types of identification and traceability systems that a facility may use.

1. Means of tracking *input* (supplied products or services) from the source all the way through the process. For example, a printer may want to be able to identify the specific source of paper used for various printing projects.
2. Means of identifying specific *operations* on the output. Operations can apply to equipment or personnel. For example, the printing facility may want to be able to trace output back to a particular press or employee.
3. Means of identifying the *ultimate destination* of output. For example, the printing facility may wish to trace the whereabouts of sensitive or important printed products, such as blank checks and stock certificates.

The extent to which a facility implements identification and traceability procedures depends on its quality needs. This is why the standard uses the term *appropriate.* In some cases, these procedures may be completely unnecessary. In other cases, they may be mandatory, i.e., required by law. Often, the need falls somewhere in between.

- Input traceability may be integral to the facility's subcontractor management system. For example, the printing facility's procedure for tracking the source of paper may be part of its ongoing evaluation of its paper suppliers.
- Operational traceability may be part of procedures to monitor employee effectiveness, equipment maintenance, or other critical process-oriented characteristics.
- Traceability to destination may be a key to ensuring long-term customer satisfaction. The most obvious example of this is in automobile manufacturing, where automakers routinely issue recalls when safety and/or quality problems arise.

When identification and traceability systems are used, the standard requires that the traced output be given a unique identification and that all such identifications be recorded.

Control of Nonconforming Product (ISO 9001: 1994, 4.13; ISO 9004-1, 14; ISO 9001 and 9004: 2000, 8.3)

Checklist of requirements:

- The facility must maintain documented procedures for the responsible segregation and disposition of nonconforming material in order to prevent its inadvertent use.

The standard's guidelines for verifying conformance to requirements were examined in Chap. 10. These verifications can be process control or product oriented. They can occur at many different points in the process:

- At the input phase
- At various within-process phases
- At the output phase
- On delivery to the customer

Inevitably, regardless of nature, these measures will result in the identification of nonconforming material of some type. The standard makes it clear that the quality system should include procedures for controlling nonconforming product to keep it from reaching customers. These procedures, according to the standard, should start as soon as the nonconformity is detected. As part of its more generic approach, ISO 9001: 2000 has renamed this element *control of nonconformity.*

ISO 9000 outlines these guidelines for controlling nonconforming product:

Responsibility. As with all other aspects of the quality system, responsibility for controlling product must be defined.

Identification. Nonconforming material should be appropriately identified. Where pertinent, other output within proximity of the nonconforming output should be examined.

Documentation. The occurrence of nonconformance should be documented at once.

Segregation. Where possible, nonconforming output should be physically segregated until disposition is made.

Evaluation. Designated and competent personnel who examine product to determine the optimal disposition must be defined.

Disposition. Output should be disposed of in accordance with documented procedures. The standard suggests four possible outcomes for nonconforming product:

- Repair to meet specified requirements, with reinspection
- Rework for a different purpose, to other specified requirements, with reinspection
- Accept with or without repair, with advice to and approval by the customer, and with a record of each occurrence maintained
- Scrap

Prevention. Steps should be taken to prevent recurrence of the nonconformity. These steps may be part of an overall corrective action process, as described below.

Handling, Storage, Packaging, Preservation, and Delivery (ISO 9001: 1994, 4.15; ISO 9001 and 9004: 2000, 7.1, 7.5.4)

Checklist of requirements:

- The facility must operate a planned, controlled, and documented system to ensure that the quality integrity of materials and output is maintained up to the time of being put into use.

As noted throughout this book, the ISO quality system is comprehensive. Other quality tools and techniques cover specific aspects of output design and production. ISO 9000 is a structured cycle of activities, beginning and ending with the customer, that governs the quality of output from conception to final delivery.

ISO 9000, then, asserts that quality measures must be extensive enough to ensure the quality integrity of output not only to the end of the process but also, if appropriate, to the point where the customer has put the output to use.

This element of the standard explains the quality system requirements that are intended to ensure quality through the delivery or installation stage. Under the standard, these requirements are part of a documented system to properly plan and control materials and output. ISO 9001: 2000 has renamed this element *preservation of product.*

The following factors are taken into account:

- *Identification* of all materials and output should be made in accordance with procedures and specifications.

- *Packaging* should be appropriate to use and environment, consistent with the contract and governed by written procedures.
- *Handling and storage* should be appropriate to the protection of the output and its shelf life and consistent with the contract. The facility should have procedures for periodic inspection of stored items.
- *Preservation* of product should be guaranteed, through segregation if necessary, while product is controlled by the supplier.
- *Delivery* should be conducted in accordance with procedures aimed at protecting the quality of product.
- *Installation* should, once again, be covered by documented procedures. These should provide for necessary instructional documents or other measures aimed at minimizing improper installation or use. Warnings as to hazards and safety should also be clearly documented.

Corrective and Preventive Actions (ISO 9001: 1994, 4.14; ISO 9004-1, 14, 15; ISO 9001 and 9004: 2000, 8.4, 8.5.2, 8.5.3)

Checklist of requirements:

- Responsibility for handling nonconformity and corrective action should be defined and documented.
- Nonconforming material should be identified and segregated until disposition.
- Nonconforming material should be reviewed and disposed of in accordance with defined procedures.
- Nonconformances should be evaluated to assess their impact on overall quality and customer satisfaction.
- Causes of nonconformances should be identified, and appropriate corrective actions should be designed and implemented.
- Where appropriate, permanent process changes should be implemented to prevent the recurrence of systemic nonconformances.
- Handling of nonconformity and corrective or preventive action activities must be documented in accordance with procedures.

By now you can see the cycle or the system created by ISO 9000 requirements. The quality system verifies conformance to requirements and makes the conformance status visible. Output that conforms goes on to the next step.

This section addresses the standard's guidelines for handling nonconforming output. These are, arguably, the most important sections of ISO 9000. The guidelines here are intended to ensure (1) that nonconforming output is prevented from reaching customers and (2) that causes of nonconformances are identified and corrected, resulting in continuous improvement.

The quality system should be structured to mesh the processing of nonconformity with corrective action. This is the best way to ensure consistent and continuous improvement at all levels. The guidelines of the standard can be expressed in the following steps:

Responsibility. As is true with every other aspect of the quality system, responsibility for handling nonconformance and taking corrective action should be assigned and documented. A specific coordinating activity for corrective action should be determined.

Identification. Obviously, all nonconformities must be identified. This usually occurs as a function of the verification activities described above.

Segregation. Where possible, nonconforming material should be segregated from conforming material to prevent it from inadvertent release and to ensure its proper disposition.

Disposition. Possibilities here include

- Reworking material to render it conforming.
- Obtaining approval from the next process step to accept as is. Usually some concession will be required from the customer. Such activity should be thoroughly documented.
- Reclassifying material to a different grade or category in which it conforms to prevailing requirements.
- Disposal.

Evaluation. The nonconformance should be studied to assess its impact on costs, performance, and customer satisfaction and to determine the facility's level of response in terms of corrective action.

Investigation. Processes, operations, records, customer complaints, and related matters should be examined to determine the root cause of the nonconformance.

Preventive action. The facility must respond to the occurrence with preventive action that is appropriate to the existing and potential impact of the nonconformance on quality and customer satisfaction. Preventive actions may entail any or all of the following:

- Process changes
- Specification changes
- Quality system changes

Process controls. Implementation of corrective action should be coupled with process controls sufficient to ensure the efficacy of the corrections and to facilitate later review and necessary adjustment. Process controls may be temporary or permanent, as appropriate.

Documentation. As with all other essential elements of the quality system, handling nonconformances and developing and implementing corrective actions should be appropriately documented.

ISO 9000 cautions that preventive and corrective actions must be more than merely effective against the specific problems. They also must be proportional to the magnitude of the problem and not harmful to associated processes.

Chapter 12

New Requirements from the 2000 Revisions

Under the 2000 revisions, ISO 9001 will be completely restructured and contain some additional requirements. Please note, however, that the revisions described herein are stilll in draft form as of this writing and are therefore subject to change.

Process-Based Structure

As mentioned in Chap. 4, ISO 9001: 2000 uses a process-based structure. The 20 elements of ISO 9001: 1994 will be replaced by five clauses containing 23 elements. ISO 9004: 2000 is organized identically. Major clauses of ISO 9001: 2000 are

4 Quality Management System (two elements), which briefly states general requirements, including such documentation as procedures and work instructions. It sets the general framework to establish a quality management system, which defines and manages processes in order to deliver a good product or service.

5 Management Responsibility (six elements), under which management defines policy, objectives, planning, and quality management system requirements while providing for feedback through management review for change authorization and initiation of improvement. It addresses management's responsibility to establish a system that continually meets customer needs and expectations, even in times of organizational change. This responsibility includes quality objectives at each organizational function and level.

6 Resource Management (four elements), where necessary resourccs, such as human resources and facilities, are determined and applied. These resources, which include new elements covering facilities and work environment, are required to implement and maintain the quality management system.

7 Product Realization (six elements), under which processes, such as customer satisfaction, design, purchasing, and production and service operations, are established and implemented. These processes are needed to manufacture product and/or deliver services from receipt to delivery. Organizations must define and describe their unique business processes, but are not obliged to use the standard's structure.

8 Measurement, Analysis, and Improvement (five elements), where results are measured, analyzed, and improved through internal audits, nonconformity control, and continual improvement. Organizations are required to measure and monitor product and/or service conformity and process and system performance. Collected data must be analyzed to initiate continual improvement.

All quality management system requirements for achieving conformity of product and/or service may be placed within this process model. For example, management defines requirements under management responsibility; necessary resources are determined and applied within resource management; processes are established and implemented under product realization; customer satisfaction and other results are measured, analyzed, and improved on through measurement, analysis, and improvement; and then management review provides feedback to management responsibility for change authorization and initiation of improvement.

While the 2000 revisions mostly contain elements already present in ISO 9001: 1994, although organized differently and expanded in some cases, there are some additional requirements. These new elements are briefly described below, except for Element 7.2.3, Customer Communication, which was previously covered in Chap. 8.

Customer Focus (ISO 9001 and 9004: 2000, 5.2)

This element requires top management to ensure that customer needs and expectations are determined, converted into requirements, and fulfilled with the aim of achieving customer satisfaction. ISO 9004 expands coverage to interested party needs and expectations. Interested parties also can include end users, people in the organization, owners and investors, suppliers and partners, and society in general.

Under ISO 9004, Element 5.2.2, an organization should define customer needs and expectations by

- Identifying and categorizing customers
- Defining the markets they compete in
- Identifying and assessing competition in their markets
- Determining key product and/or service features and their relative value for customers
- Identifying opportunities, weaknesses, and future competitive advantages

Internal Communication (ISO 9001 and 9004: 2000, 5.5.4)

This element requires the organization to ensure communication between the various levels and functions regarding the quality management system and its effectiveness. ISO 9004 states that communication tools may include

- Team briefings and other meetings
- Notice boards and in-house journals and magazines
- Audiovisual and electronic media

Assignment of Personnel (ISO 9001 and 9004: 2000, 6.2.1)

This element requires the organization to ensure that people with quality management system responsibilities are competent on the basis of applicable education, training, skills, and experience. ISO 9004 suggests that to achieve its objectives and stimulate innovation, an organization should encourage the involvement of people through

- Identifying competence needs
- Selection, ongoing training, and career planning
- Managing performance toward its objectives by establishing individual and team objectives and evaluating the results
- Facilitating involvement in objective setting and decision making
- Encouraging recognition and rewards
- Facilitating open two-way communication
- Creating conditions to encourage innovation
- Ensuring effective teamwork
- Using information technology to facilitate communication of suggestions and opinions

- Using people satisfaction measurements for improvement
- Investigating why people leave the organization

Facilities (ISO 9001 and 9004: 2000, 6.3)

This element requires the organization to identify, provide, and maintain the facilities it needs to achieve product conformity. These facilities include workspace and associated facilities; equipment, hardware, and software; and supporting services.

ISO 9004 states that the organization should define and provide an infrastructure specified in such terms as objectives, function, performance, availability, cost, safety, and security. There should be a process to develop and implement a maintenance program and strategies to maintain the quality of product and/or service. The organization should evaluate the infrastructure against the requirements of all interested parties, giving consideration to environmental issues, and plan for risks from naturally occurring phenomena.

Work Environment (ISO 9001 and 9004: 2000, 6.4)

This element requires the organization to identify and manage the human and physical factors of the work environment needed to achieve product conformity. ISO 9004 states that work environment factors influence employee motivation, satisfaction, development, and performance; product and service quality; and the well-being of people and their ability to contribute to achieving organizational objectives. As such, the quality management system should ensure that the work environment supports achievement of organizational policies and objectives.

Human factors that may affect the work environment include

- Creative work methods and opportunities for greater employee involvement
- Safety rules and procedures, including protective equipment
- Ergonomics
- Special facilities for people in the organization

Physical factors that may affect the work environment include noise, heat, light, hygiene, humidity, cleanliness, vibration, pollution, and airflow.

Identification of Customer Requirements (ISO 9001 and 9004: 2000, 7.2.1)

This element requires the organization to determine customer requirements, including the customer's specified product requirements, such as those for availability, delivery, and support; product requirements not specified by the customer but necessary for intended or specified use; and obligations related to product, including regulatory and legal requirements.

ISO 9004 states that the organization should seek and review relevant information to ensure understanding of customer and other interested party needs and expectations. It should define and implement the processes for establishing these requirements and then communicate it throughout the organization and to customers and other relevant interested parties.

Measurement and Monitoring (ISO 9001 and 9004: 2000, 8.2)

This element was examined in part earlier. Element 8.2.2, Internal Audit, was covered in Chap. 8, while Element 8.2.4, Measurement and Monitoring of Product, was described in Chap. 10. The remaining two elements are examined below.

Element 8.2.1, Customer Satisfaction, requires the organization to monitor information on customer satisfaction as a measurement of quality system performance while determining the methodologies for obtaining and using this information. ISO 9004 states that the organization should identify sources of customer-related information and establish processes to gather, analyze, and use it. The organization's system for determining and monitoring customer satisfaction feedback should address aspects of product and/or service quality, price, and delivery and should be provided on a continual basis.

The organization should cooperate with its customers in order to anticipate future needs, specify the methodology and measures to be used and the frequency of review, and plan appropriate data collection methods.

Element 8.2.3, Measurement and Monitoring of Processes, requires the organization to apply suitable methods to measure and monitor the processes necessary to meet customer requirements. These methods must confirm the continuing ability of each process to satisfy its intended purpose.

ISO 9004 states that the organization should identify measurement methodologies and perform measurements to evaluate process performance, how the measurements are incorporated into the product realization process, and the role of measurement in process management.

Part

3

Putting ISO 9000 to Work

Chapter

13

Implementing ISO 9000

Suppose you want to put ISO 9000 to work in your facility. Perhaps you have a shot at a major new account, but ISO 9000 registration is required before the client will even talk to you. ("Don't even bother with a proposal until you're registered.")

Maybe your biggest customer has told you that your main competitor has become ISO 9000 registered. ("People over there claim it's improved their quality 300 percent. What do you think?") Or even more traumatic, your primary customer is requiring ISO 9000 registration of its entire supplier network. ("We're cutting our supplier group by half. Anybody not ISO registered is out.") Or maybe your facility has no formal quality system now, but management has decided that a quality system is a key to maintaining and expanding markets.

Implementing an ISO 9000 quality system is no different from making any other major, fundamental, and far-reaching change in your organization. In plain terms, how difficult is implementing ISO 9000? The answer depends on many factors. The amount of pain, expense, and time required to implement ISO 9000 depends on

- The sophistication of your existing quality program
- The size of your facility
- The complexity of your process

Implementing an ISO 9000 quality system is, in fact, as simple as

- Knowing the requirements of the standard
- Understanding the facility and its processes
- Committing the resources

- Planning well
- Seeing it through

Add to both these lists one other factor, in fact, the biggest one of them all—management commitment. Without it, all else is for naught. "Top management needs to know a lot about ISO 9000," says Jim Ecklein of Augustine Medical. "We gave our managers a lot of training in it, got them walking, talking, breathing ISO."

If you are a career quality professional, all this probably sounds very familiar to you. No surprise. In terms of implementation, ISO 9000 is not much different from any other quality tool or technique. Same tough sell. Same painful transition period. Same frustrations. Same profound impact.

ISO 9000 implementation differs from that of other quality programs in that it affects the entire organization. When faithfully and aggressively pursued, it results in that always coveted but seldom realized cultural transition to an atmosphere of continuous improvement. There are other advantages too, such as

- Improved customer satisfaction
- Increased competitive strength
- Greater market share
- All those other benefits described in Chap. 3

However, it is virtually impossible to get there until you undergo a formal ISO 9000 implementation cycle. Take the word of one who tried. "We wrote an ISO 9000 quality manual," says Don Van Hook of Strahman Valves, "but the consultant who came in and assessed it said it needed work; no way would we be approved." With the help of the consultant, Strahman implemented an ISO 9000 quality system and was registered in 1995.

In fact, according to a report in *Compliance Engineering,* 70 percent of all companies that "do not pursue preassessment or preparation services" (implementation consulting) fail the ISO 9000 preassessment the first time through.

This chapter examines the steps of ISO 9000 implementation and the importance of consultant assistance. Some of the observations here are general enough to apply to quality improvement programs in general. Others are specific to ISO 9000. The main focus is on creating an implementation process that (1) gets an ISO 9000 quality system up and running and (2) prepares the facility for ISO 9000 registration, if desired. The registration process is the subject of Chap. 14.

The Absolute Prerequisite: Top Management Commitment

Deming said it. Juran said it. Crosby, Feigenbaum, and Peters said it. Every quality consultant says it, and most company managers concur. Without it, no quality initiative can succeed. With it, no quality initiative, including ISO 9000 implementation, can fail.

Top management commitment is that important.

Ireland's Lorcan Mooney, an ISO 9000 consultant, implementer, and quality systems auditor, says: "The most significant pitfall in ISO implementation is the CEO being uncommitted, or opting out, or standing on the sidelines and expecting other people to do it."

Management responsibility guidelines come first in both the old and new versions of the standard because their creators understood that without management involvement and participation, nothing fruitful can happen.

So it is with ISO 9000 implementation. In this context, commitment is much more than approving purchase orders and observing pep rallies. The CEO cannot delegate commitment, passion, and resolve. The CEO cannot remain passive, let alone skeptical, and expect to see results. It is said that 80 percent of success is showing up. With ISO 9000 implementation, the figure is more like 100 percent. *Top management must show up.*

What is more, top management has to be an active, daily, and visible presence in the process. Brian Burke, of Container Products Corporation, observes: "Top management people have to understand and do more than just accept the tenets of the ISO 9000 process. They have to lead it. They don't necessarily have to get involved in the nitty-gritty, but they have to lead and drive and apply pressure when necessary."

Where does this type of top management/CEO commitment come from? There seem to be several sources:

- Direct marketplace pressure: requirements of crucial customers or parent conglomerates
- Indirect marketplace pressure: increased quality levels and visibility among competitors
- Growth ambitions: desire to exploit European Union market opportunities
- Personal belief in the value of quality as a goal and quality systems as a means of reaching that goal

One of these forces is usually enough to inspire top management interest in ISO 9000. True commitment usually springs from some combination of the preceding. It is not something that can be imposed 100 percent on the CEO. Much of it has to come from genuine conviction.

However, once top management commitment is there, CEOs make ISO 9000 implementation, and operation of the quality system, a top corporate priority, and make sure that managers never forget it. This is certainly the case at Container Products, where, according to Brian Burke,

> The CEO chairs quarterly meetings of our top plant managers. At each meeting, plant managers present various reports on key business areas: financial, operations and sales. Usually, since our CEO is a finance type, the first items reported upon are the financials. When we started implementing ISO 9002, I suggested to the CEO that he have managers present status reports on ISO 9002 implementation first, even before the financials. That's what he did. It caught the plant managers off guard, but it spoke loud and clear as to the priorities of the organization.

A quality professional whose firm has become registered to ISO 9000 relates another example of CEO firmness:

> One of VPs walked into his office and told the CEO that ISO 9000 was never going to work here, and that he didn't want to be involved. The president's answer was, "Either sign on or you're out of here." Needless to say, the VP signed on. So did everyone else.

Contrast this with the experience of another quality professional, struggling to implement ISO 9000 in an automotive manufacturing plant in the Northeast:

> You've *got* to have management commitment behind you. When we started out, they told me they were committed, but now I know they aren't. How did I find out? When I called some meetings to discuss ISO, and no one showed up.

One would charitably have to call this ISO 9000 implementation program an uphill battle.

The Role of the Consultant

On average, it takes approximately 18 months to implement ISO 9000, at a cost of $135,000 for a 100-person facility. Registration adds to this amount.

At Perry Johnson, Inc. (PJI), we have tried many different ISO 9000 implementation approaches over the years. We found in our experience that it is impossible to train client employees to properly write ISO 9000 documentation in 1 week. After training was completed, it usually took the employees 1 to 2 years to write the documentation, with a resulting quality system that often was cumbersome to maintain. Clients frequently failed to understand the standard and made unnecessary, expensive, and time-consuming operational changes in attempts to fulfill its requirements. These actions sometimes had negative effects on product or service quality.

We have now learned that the most efficient, reliable, and cost-effective way to implement ISO 9000 is to have consultants prepare the quality manual and procedures while concurrently assisting in implementation and providing training to client employees. PJI follows this approach in its patented A to Z Complete Implementation Program, which can cut a client's costs by about 80 percent and enable it to register in only 6 months.

Writing documentation is a very delicate process. Whatever is documented must be followed on a consistent and provable basis. The more that is written, the greater is the implementation burden on the company, and the greater is its potential nonconformance exposure during audits. Therefore, the consultants must only write what is necessary to meet ISO 9000 requirements. This approach will be further facilitated by ISO 9001: 2000, which can be tailored to match an organization's operations.

The resulting documentation should

- Accommodate the standard to fit the client's processes
- Minimize operational changes, paperwork, and maintenance costs
- Maximize the client's ability to pass the registration audit

The consultants are the key to successful implementation, which requires significant expertise in both ISO 9000 and documentation writing. PJI uses a team approach, with an experienced lead consultant and one or two others, for it usually takes six implementation efforts before a consultant can become a group leader.

Typically, the consultants visit a client's facility for 1 week. During that time, they gather the necessary information, review the quality manual and procedures (if any), conduct a benchmark verification audit to discover gaps between the client's quality system and ISO 9000 requirements, and provide a brief overview to management.

Over a 90-day period, they write the quality manual and procedures; assist in implementing the ISO 9000 quality system, selecting the management representative, conducting internal audits, and taking corrective action; act as liaison with the registrar; and train the management representative, along with 5 to 10 percent of the client's workforce, primarily in internal auditing, an ISO 9000 requirement.

The system must be in place for at least 90 days, in order to generate a sufficient paper trail, before a registration audit can be conducted. PJI recommends that at least one client employee attend an ISO 9000 lead auditor course in order to understand how a registrar auditor will evaluate and grade the facility. PJI consultants conduct a mock audit of the quality system before the registration audit, follow up with corrective actions, and then provide consulting services during the registration audit, correcting nonconformances on site in order to speed up the registration process.

Establish Implementation Teams

ISO 9000 is implemented by people, not by magic. The first phase of implementation requires the commitment of top management, the CEO, and perhaps a handful of other key people. The next step, even with consultant assistance, is creating a personnel structure to plan and oversee implementation.

The first component of this personnel structure, and the most important one after the CEO, is the *management representative* (MR). In the context of the standard, the MR is the person within the facility who acts as liaison between facility management and the ISO 9000 registrar.

However, the role is much broader than this. The MR also should act as the facility's quality system champion, the protector of the vision. The MR must be a person with

- Access to and total backing of the CEO
- Genuine and passionate commitment to quality in general and the ISO 9000 quality system in particular
- The clout, resulting from rank, seniority, or both, to influence managers and others at all levels and functions
- Detailed knowledge of quality methods in general and ISO 9000 in particular

The MR should be a person who enjoys the trust, confidence, and total backing of the CEO. He or she should also have the respect of the other facility employees.

Next, a top-level implementation team is created. In some firms, this is called the *quality action council.* In others, it is known as the *quality steering committee.* This team usually is chaired by the MR and consists of

- The CEO
- Top managers
- Key functional managers
- Top union representative (if applicable)

The quality action council is mainly a policy group. It sets objectives for quality system implementation, approves plans, evaluates reports, and prescribes changes as needed. The council also makes critical decisions about the quality system documentation. Its members should decide early on who is responsible for (1) writing, editing, and approving the facility's quality manual and (2) second-tier and (if used) third-tier documentation.

Next is a network of *quality action teams.* In small to medium-sized facilities, these teams are organized by department or function. Each is headed by the functional manager or department head, who in turn sits on the quality action council. Larger facilities have more elaborate action team networks, but the objective is the same: to have a team of line people representing each critical facility function and process element. Whatever organization is devised, there must be a clear and visible reporting and communication network, all the way up to the MR and the quality action council.

All members of the quality action teams are knowledgeable about the process elements in which they work. This is the main criterion for membership. It is helpful, though not required, that they also have some advanced familiarity with ISO 9000 quality systems. However, such familiarity can be provided by an ensuing training cycle.

The quality action teams are the hands-on element of the ISO 9000 implementation effort. They execute the policy devised by the council. They make the quality system happen.

There is one other people issue: Should facilities engage the services of an outside consultant? As noted earlier, most find that this step pays major dividends in reducing implementation time and expense and increasing effectiveness. Larger organizations with multiple facilities to implement often employ a full-time specialist to help them get their ISO 9000 quality systems up and running.

Assess Current Quality System Status

As we have seen in our review of the ISO 9000 technical requirements, there is little that is novel about the quality system. Most of its elements are a combination of good common sense and the best of quality methods, techniques, and philosophies that have been around for years.

Keep this in mind as you begin the implementation process. There is nothing in the ISO 9000 standard that requires duplication of effort, redundant systems, or make-work. To the contrary, the goal of ISO 9000 is to create a quality system that conforms to the standard. This does not preclude incorporating, adapting, and adding onto quality programs already in place. In fact, the standard encourages facilities to do just that.

Therefore, the next step in the implementation process is to compare the facility's existing quality programs and quality system, if there is one, with the standard's requirements. Program assessment can be done internally, if the knowledge level is there. Otherwise, a benchmark verification audit or gap analysis can be conducted by any one of a large number of ISO 9000 consulting, implementing, and registration firms.

Facilities already subject to customer quality audits often find that their existing system meets a sizable number of the ISO requirements. This is especially true of facilities conforming to the various military standards.

Dennis Beckley of Dayton Rogers found that MIL STD 45208 and 45662 had many elements found in ISO. David Turteltaub of Phillips Circuit Assembly, which conforms to MIL-Q-9858A, says that that standard "is parallel to ISO in many ways. The intent is the same, except for two or three different areas. They share many of the section titles, groupings and wording of the various subsections."

The same is true for customer-mandated quality standards such as QS-9000, the automotive Big 3's suppliers quality program. QS-9000 is a harmonized version of ISO 9001 containing the same 20 elements. However, these elements also contain additional sector-specific requirements that are not present in ISO 9001, and there are additional customer-specific requirements. As will be demonstrated in Part 4, some customer-mandated quality systems have become ISO 9000 derivatives.

Facilities that have subscribed to Malcolm Baldrige National Quality Award criteria also find their ISO 9000 implementation path easier than others. "GE Automation," according to *Managing Automation* magazine, "had an easier time than many in winning ISO 9001 registration…because the company had previously applied for the Malcolm Baldrige National Quality Award. It already had extensive outside comment on its quality procedures."

Facilities with formal quality systems already in place are well on their way toward implementing full-fledged ISO 9000 quality systems. For those without quality systems, implementation can be quite an uphill battle. Sandy Weller of Woodbridge Foam Corporation is responsible for implementing an ISO 9002 quality system in a facility that had no quality systems at all. Her assessment? "It's no fun."

If not undertaken by a consultant, the benchmark verification audit is conducted at several levels. The quality action council focuses on the larger elements of the ISO 9000 standard, including management responsibility, quality system, quality costs, quality audits, and personnel and training (see Chap. 5). Quality action teams compare the standard's requirements with systems and procedures in their own functional areas.

The main purpose of the benchmark verification audit or gap analysis is to get a clear picture of the state of the facility's quality program as it compares with ISO 9000 requirements. The consultant or the teams should pinpoint systems that conform, systems that can be adapted, and most critically, areas of nonconformance.

As with so many other aspects of ISO 9000 implementation, the gap analysis process often can be beneficial on its own merits. At Adhesives Research, department members assessed one another. "That gave all of us a chance to vent our venom," says Anthony Coggeshall. "It didn't hurt the process. It actually helped it, because we got all that good and bad and peripheral garbage on the table, where we could sort through it, and fix it up, and prioritize it."

Information gathered in the benchmark verification audit or self-assessment should be documented and channeled to the quality action council. If the assessment is conducted with the help of a consultant or other outside body, that organization will usually provide specific implementation direction at the same time.

Create a Documented Implementation Plan

Once the facility has obtained a clear picture of how its quality system compares with the ISO 9000 standard, all nonconformances must be addressed by a documented implementation plan. This plan may be created by an ad hoc committee under the authority of the quality action council, or it may be prepared by a consultant. Usually, the plan calls for setting up procedures to make the facility's quality system fully conform to the standard. If not written by a consultant, procedures that affect high-level policy elements of the quality system may be handled by the council itself or designated members. Others may be handed down to various quality action teams for development.

The implementation plan should be thorough and specific, detailing

- Procedures to be developed
- Objectives of the system
- Pertinent ISO 9000 elements
- Person or team responsible
- Approval required
- Training required
- Resources required
- Estimated completion dates

These matters should be organized into a detailed GANTT or timeline chart to be reviewed and approved by the quality action council. Once approved, the plan and its GANTT chart should be controlled by the MR. The chart should be reviewed and updated at each council meeting as the implementation process proceeds.

As mentioned earlier, quality system areas requiring implementation can vary widely between facilities and depend on the type and level of quality system elements already in place. Some facilities, with long-standing total quality management (TQM) or other quality systems, may need only rudimentary implementation measures to deal with nonconformances. Others may require the development of full-blown quality systems, a process that can take months, if not years.

Most facilities with some level of quality program already in place find that their biggest areas of nonconformance are in

- Design (for ISO 9001)
- Purchasing
- Inspection and testing
- Process control

Dennis Beckley of Dayton-Rogers found that management responsibility was a weak area in the quality system. "We have had good management commitment," he says. "We've been operating a TQM program for 4 years. We have mission statements and so forth, but management responsibility was not an area addressed in our manual or quality system per se."

Another potential problem area is training. Most facilities have training programs of some sort, and many programs are in fact very good. Where they sometimes fall short of ISO requirements is in the needs-assessment area. "Our training program was basically reactive," notes Brian Burke of Container Products. "When training was necessary, we did it. We didn't follow ISO's proactive approach of constantly evaluating training needs to see what we should be doing."

Among American firms, the most notorious area for nonconformance is *documentation*. This will be described in more detail later.

Provide Training

Obviously, the implementation plan will have provided for training in various functional areas of the quality system. Training *must* be provided to all employees performing activities that can affect quality. Certain training needs will depend on the nonconformances being addressed. The quality action teams should take responsibility for providing specific functional training in their functional areas.

In addition, more general training needs apply to implementation and virtually all facilities. Since the ISO 9000 quality system affects all areas and all personnel in the organization, it is wise to provide basic orientation in the quality system standard to all employees. This can be a 1-day program that educates personnel about quality systems in general and the ISO 9000 quality system in particular. Consultants can conduct the training.

Adhesives Research had its own program. "We set up a formal ISO 9000 training syllabus for management, supervisory and hourly people," says Anthony Coggeshall. "It was not lengthy, just a good overview, supported by transparencies, covering basically what ISO 9000 was and, more importantly, what it was not."

The training program should emphasize the benefits that the facility expects to realize through its ISO 9000 quality system. The program also should stress the higher levels of participation and self-direction that the quality system offers to employees. Such a focus will go far to enlist employee support and commitment.

"We had no problem getting our hourly people on board," says Sandy Weller of Woodbridge Form Corp. "They were really happy to be getting some say. That seems to have been the big thing, that our people were getting some say."

David Turteltaub of Phillips Circuit Assembly agrees. "Our people were very favorable. They liked to see the structure being implemented to help them out in their jobs. They also liked the uniform approach to the different areas of the facility. And it helped for them to know that the facility management group was solidly behind the program, from the president and the general manager level on down."

If a consultant does not write procedures, the quality action council should create and distribute a standard covering procedure writing for the facility. This will ensure that all procedures, the heart of the quality system's documentation, are consistent and relevant.

Create Documentation

As noted earlier, documentation is the most common area of nonconformance among American facilities implementing ISO 9000 quality systems. As Don Van Hook of Strahman Valves observes: "When we started our implementation, we found that documentation was inadequate. Even absent, in some areas. Take calibration. Obviously it's necessary, and obviously we do it, but it wasn't being documented. Another area was inspection and testing. We inspect and test practically every item that leaves here, but our documentation was inadequate."

There is no way around it: Documentation is mandatory. It is essential to the ISO 9000 registration process because it provides objective/audit evidence of quality system status.

The three basic rules of ISO 9000 documentation are

- Document what you do.
- Do what you document.
- Prove it.

Many facilities find that their existing documentation is adequate in most respects. To bring it into full ISO conformance, they implement control procedures to ensure that documentation is available as needed, and is reviewed, updated, stored, and disposed of in a planned and orderly manner.

In this context, documentation mainly refers to procedures. Under the ISO 9000 quality system, all work that affects quality must be planned, controlled, and documented so that it can be performed in a way that maximizes quality. Toward this end, detailed procedures and work instructions must be created where, as the standard states, their absence would adversely affect quality.

For many facilities, the process can become nightmarish. Don Van Hook observes: "One of hardest things to do is to get shop people to become what they think of as bookkeepers. They don't like paperwork. To them it's a dirty word."

But the process of creating and using documentation is central to the effectiveness of quality system implementation. For one thing, the exercise is almost always educational. Many facilities that find that creating work procedures and instructions is a difficult process are facilities that have never had such things before. Typically, important process work, even work with a serious impact on quality, is done in an unplanned and inconsistent manner.

Quality system documentation forces facility personnel, at all levels, to think through exactly what is being done and how it is accomplished. Most firms find this to be a positive learning experience, for it

- Disseminates knowledge of the process
- Pools process knowledge and expertise
- Creates positive interactions among individuals and process elements

Through these interactions, people develop positive and proactive approaches to teamwork and quality as they

- Learn how to work together better
- Learn what to expect of each other

- Establish communication channels that result in positive improvement

This is far more powerful, and far more important, than the mere process of writing down procedures and work instructions. Writing them down, however, is a vital part of the documentation process, too. Written documentation is

- Evidence that thought has been given to the procedures
- An irreplaceable reference resource for outside auditors
- An invaluable training and improvement tool

Keep in mind also that quality system documentation does not need to be exhaustive, exhausting, or redundant. It is certainly not intended to be an end in itself. Again, ISO 9000 requires only that the facility have such documentation as procedures and work instructions where their absence would adversely affect quality. The facility must maintain the minimal amount of documentation necessary to demonstrate that its quality systems exist and are being operated.

As you have probably gathered, the process of creating the needed documentation can be arduous. While it was stated earlier that consultants should write the procedures with client employee input, facilities should make sure that work instructions are created by the people who actually do the work.

Sandy Weller of Woodbridge Foam says that her facility went about the task another way: "At first, I wrote the procedures, not the people doing the work. So when we put the procedures into practice, the people doing the work had some problems. But we're working out the bugs."

Dennis Beckley of Dayton-Rogers had a similar experience. He and other corporate implementers played a major role in creating quality documentation and manuals for the various facilities implementing ISO 9000. "We tailored the documentation to one plant's way of doing things. At our corporate quality meetings, plant managers reviewed the documentation and approved it. Then we audited the plants and found out they were doing things their own way and not following the procedures. From a corporate standpoint, it can be tough to get everyone to do what's in the documentation."

Even though work instructions and possibly procedures should be documented by the people actually doing the work, associated middle managers should be involved in the process as well. The experience of Container Products demonstrates the importance of this involvement. According to Brian Burke, in most Container Products plants the quality assurance people worked directly with the line people on the procedures. "Even though the supervisors and product managers signed off on the procedures, they were not nearly as enthusiastic about implementing them as were the supervisors in the plant, where everyone was included in the process."

The final link in the documentation chain is the facility's *quality manual.* The contents of the manual were examined in detail in Chap. 6. From an implementation standpoint, quality manual creation is driven by the top-level quality action council or the consultant. They must draw on the systems, procedures, and documentation created and implemented by the quality action teams.

The quality manual must address, point by point, the components of the ISO 9000 quality system being implemented. Ultimately, the quality manual is

- The blueprint for the facility's quality system
- The cross-reference of the facility's quality procedures
- The expression of the facility's quality commitment

Document Control

Once the necessary quality system documentation has been generated, a documented system must be created to control it. As noted in Part 2, control is simply a means of managing the creation, approval, distribution, revision, storage, and disposal of the various types of documentation. Document control systems should be as simple and easy to operate as possible and need only be sufficient to meet ISO requirements.

The principle of ISO 9000 document control is that employees should have access to the documentation and records needed to fulfill their responsibilities. Ironically, direct access often can result in certain employees having *less* recordkeeping and documentation to deal with and can be a cause of resistance. The facility's quality manual is a prime example. "We got minor resistance from some major players who were used to having the quality manual, but who didn't really need to have their own copy of it," says Jim Ecklein of Augustine Medical. "We solved that by having a master quality manual, with references to submanuals for each facility area. That way, people had what they needed, but we weren't passing quality manuals out to people who didn't really need them and wouldn't use them."

Monitor Progress

When the procedures have been completed and the quality system fleshed out, it is time to put the quality system into effect. In this extremely important phase, management must pay close attention to results to make sure that the elements of the quality system are logical and effective.

Effective monitoring is what makes or breaks ISO 9000 implementation. It is also the ultimate measure of how well, or poorly, facility management is living up to its responsibilities, as described in the management responsibility element of the standard. ISO consultant Lorcan Mooney says, "A serious threat to effective ISO 9000 implementation is failure to monitor development as the process proceeds."

In particular, management at all levels should keep an eye out for

- Gaps and assumptions in procedures
- Steps that are difficult, ineffective, or impractical

Many such problems can be dealt with by the quality action teams. Resulting changes, of course, should be documented and approved in accordance with quality system procedures.

Management, up to the level of the quality action council, simultaneously should carry out its review functions, as prescribed by the standard and its own documented procedures. These activities include

- Internal audits
- Corrective and preventive actions
- Management reviews

It is especially helpful to begin the internal audit program immediately as part of implementation, thereby solving many problems at the local level. "We have internal audits every 4 weeks," says Anthony Coggeshall of Adhesives Research. "Each team consists of members of two unrelated departments, not the quality department, and not the department being audited. These audits are really forums for departments to talk to one another and solve their own problems."

Dennis Beckley's organization, Dayton-Rogers, follows a similar practice, but on a multiple-facility basis. "We have ISO champions in each plant, and each is going to our other plants to conduct internal audits in accordance with the internal audit plan we created. This includes corrective action and follow-up as needed."

Inevitably, it will become evident that certain individuals in the facility are unenthusiastic about effective quality system implementation. Outright resistance or obstruction is, of course, a disciplinary matter beyond the scope of this book. How should management handle reluctant participants? Should it make a special effort to win them over?

No, says Lorcan Mooney. "Focus your efforts on the people who are already sold on ISO," he advises. "Don't concentrate on winning over the lukewarm or the uncommitted. Give it time. Sometimes, peer pressure, seeing coworkers thrive under the system, does the trick. In other cases, the quality system itself does the convincing. When you get a few wins under your belt, a lot of those people will fall off the fence on your side."

Review: Pitfalls to Effective Implementation

Here is a brief checklist of the most significant barriers to effective ISO 9000 quality system implementation:

1. *Lack of CEO commitment.* As Lorcan Mooney says, "If senior management consists of four or five people, and two of them are not committed, over time they can be won over. But if the CEO is not committed, then in no way are you going to win in the long run."
2. *Failure to involve everyone in the process.* Ownership and empowerment are key to effective implementation. To make employees feel like owners of their activity, make them responsible for developing and documenting their procedures.
3. *Failure to monitor progress and enforce deadlines.*

Notice that all three of these pitfalls are directly traceable to management, or lack of it.

Positive Outcomes

When effectively implemented, ISO 9000 quality systems provide positive benefits almost at once. "Of course we started with nothing," says Sandy Weller of Woodbridge Foam. "But almost from the start, we noticed a major reduction in errors. Just the process of putting in the system, documenting what we were doing, and keeping track of what was going on improved quality, because it made everyone aware of his or her impact on quality."

Chapter

14

Registering to ISO 9000

"In 15 months we went from nothing to ISO 9001 certification. I've done this before and you have to have a systems approach. A lot of quality managers don't realize that you have to give people a philosophy, a way of thinking, to get the ball rolling."

SHERMAN MCDONALD
GSE, Inc.
ISO 9001 registered

A facility registers or becomes certified to ISO 9000 in order to obtain objective third-party verification that its quality system conforms to the ISO 9000 standard. The facility may go through the registration process simply to reinforce its own quality program and commitment. There are other reasons as well. Some facilities register

- As a proactive step to counter competition and secure new business, especially in the European Union
- In response to a competitive threat
- In response to customer requirements
- In response to the requirements of a parent organization, as is often the case among multinationals

Finally, some firms, very few at this point, but the number may grow, register because they are obliged to by legal and/or regulatory authority.

The registration process itself is rather simple. Compared with creating and implementing an ISO 9000 quality system, registration may seem almost anticlimactic. Most American facilities today find that the biggest problem with registration is getting past the obstacles, confusion, and misinformation that are rife in the marketplace.

This chapter explores the ISO 9000 registration process in detail. It attempts to clear away the obstacles, sort out the confusion, and replace bad information with good. The information presented here is considered to be accurate as of press time. For the best help, however, prudent managers will seek the counsel of recognized ISO 9000 consultants.

As indicated in earlier chapters, some facilities rush into the process of applying for ISO 9000 registration before they are ready. This explains why some 70 percent of initial ISO 9000 registration applications fail on the first go-round.

Before applying for ISO 9000 registration, it is imperative that your facility has

- Implemented a quality system that meets the technical requirements of the part of the ISO 9000 quality system standard to which registration is sought, whether ISO 9001, 9002, or 9003
- Documented it with a quality manual and subordinate procedures, work instructions, and other materials
- Operated it successfully for 3, or preferably 6, months
- Conducted at least one management review

These are the preparatory steps that were detailed earlier in this book. Once they are accomplished, the typical facility takes the following steps to obtain ISO 9000 registration:

- Establish a relationship with an ISO 9000 registrar whose scope of accreditation covers the facility's field of activity.
- File a formal application.
- Undergo a documentation audit.
- Undergo an on-site quality system audit.
- Remedy nonconformances by completing any required corrective action requests.
- Pay the associated fees.
- Maintain the system.

The first step is by far the trickiest.

Hiring an ISO 9000 Registrar

According to The McGraw-Hill Companies, there are hundreds of registration firms operating in the United States today. How does a facility go about recruiting an ISO 9000 registrar?

Very carefully.

Finding a registrar is not at all tough. It is as easy as looking in the Yellow Pages. However, especially in the United States, all registrars are not created equal. All can provide some sort of registration. But not all can give you the type of registration that will help you meet your goals.

Why is this so? The short answer: The United States does not have a government-sanctioned accreditation body. To understand what this means, let us take a look at the European sphere for a moment.

The registration process overseas

In Great Britain, for example, the responsibilities for ISO 9000 are well defined, understood, and recognized.

- The top-level authority involved in quality systems registration to ISO 9000 and BS 5750 is the British government's Department of Trade and Industry (DT&I).
- Since 1985, DT&I has sponsored the United Kingdom Accreditation Service (UKAS), which accredits, in effect provides a stamp of approval on, quality system registrars.
- These registrars conform to the requirements of UKAS and are authorized to issue a registration stamp bearing the UKAS logo.
- Each registrar is accredited to provide quality system registration in certain specified vertical markets. The range of markets that registrars operate in is referred to collectively as their *scope* (see Fig. 14-1).

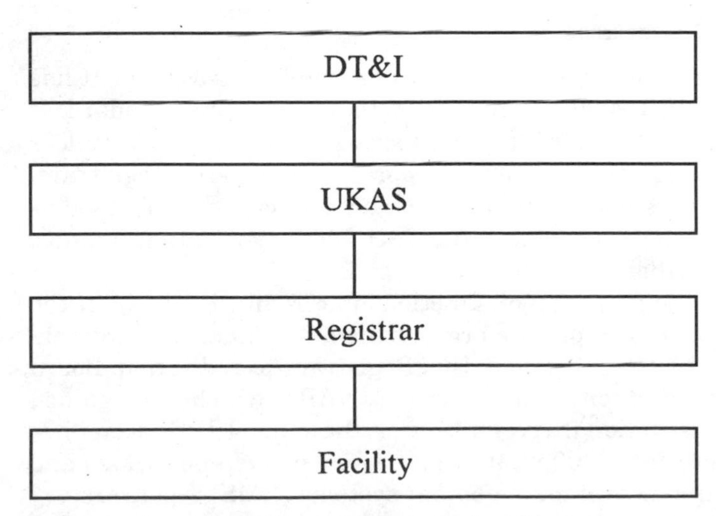

Figure 14-1. Registration hierarchy in the United Kingdom.

Thus, if you manage a British facility that makes concrete pipe, finding a registrar is relatively simple. Various publications list the UKAS-accredited registrars. You simply look up a registrar whose scope includes concrete pipe, make a call, and go from there.

Most European Union (EU) nations have government-sanctioned accreditation bodies equivalent to UKAS. These bodies, which conform to requirements published by the EU's European Union for Quality System Assessment and Certification (CQS), include the Raad voor Accreditatie (RvA) of the Netherlands, the world's most recognized accreditation body. Some extend reciprocal recognition to other accrediting bodies, whereas some do not.

There is a separate training and accreditation system for ISO 9000 auditors. ISO publishes qualification criteria for quality systems auditors in ISO 10011, which will be replaced by ISO 19011. The International Register of Certificated Auditors (IRCA) in Great Britain is the EU body empowered to sanction and approve auditor training and certification schemes.

What this Byzantine bureaucracy amounts to is that a firm wishing to obtain ISO 9000 registration that is recognized and accepted in the EU should deal with a registrar that

- Is accredited by UKAS or an equivalent body
- Uses auditors whose credentials are traceable to IRCA

As another example, registrars doing business in Japan seek accreditation from the Japan Accreditation Board for Conformity Assessment (JAB) and use auditors certified by the Japanese Registration of Certificated Auditors (JRCA).

Registration for U.S. facilities

Unlike Great Britain and other EU nations, the U.S. government has not endorsed an ISO 9000 or equivalent quality systems standard. It does not sponsor an accrediting body such as UKAS or IRCA. It does not provide recognition of or support to domestic registrars. Essentially, any organization in the United States can identify itself as a quality system registrar and issue ISO 9000 registration stamps, even if its only authority is granted by itself.

While there is no government-sanctioned accreditation body in the United States, there is a private organization, the American National Standards Institute (ANSI) and the Registrar Accreditation Board (RAB) National Accreditation Program (NAP), which has gained national and international recognition as the official U.S. accreditation body for both ISO 9000 quality management systems (QMS) and ISO 14000 environmental management systems (EMS) registrars.

As mentioned earlier, ANSI is the American ISO affiliate. RAB, formed in 1989, is a separately incorporated affiliate of the American Society for Quality (ASQ), a professional society that is an ANSI-accredited standards writing body. Both RAB and ANSI are nonprofit organizations, and they cosponsor the American equivalent to ISO 9000, ANSI/ASQ Q9000.

ANSI-RAB NAP is administered by a Joint Oversight Board (JOB) that oversees the RAB QMS Council and the ANSI EMS Council. There is equal representation of ANSI and RAB appointees on this board.

Under the ANSI-RAB NAP, RAB operates accreditation programs for ISO 9000 and ISO 14000 registrars and auditor training course providers. RAB accepts and processes applications for accreditation and then forms audit teams to conduct accreditation audits of candidate registrars and course providers. The appropriate councils review RAB audit team reports and vote to grant or deny accreditation to registrars and course providers. RAB separately administers QMS and EMS auditor certification services.

ANSI promotes the NAP nationally and internationally, provides due process and public review of all program criteria and procedures, and is responsible for public notice of applicants for accreditation. ANSI, in consultation with RAB, represents the NAP in international accreditation activities (see Fig. 14-2).

RAB and ANSI have worked together since 1991 to accredit ISO 9000 registrars. They formed the NAP in 1996 to add ISO 14000 accreditation and to allow for expansion into new accreditation program areas should additional management standards be developed. As

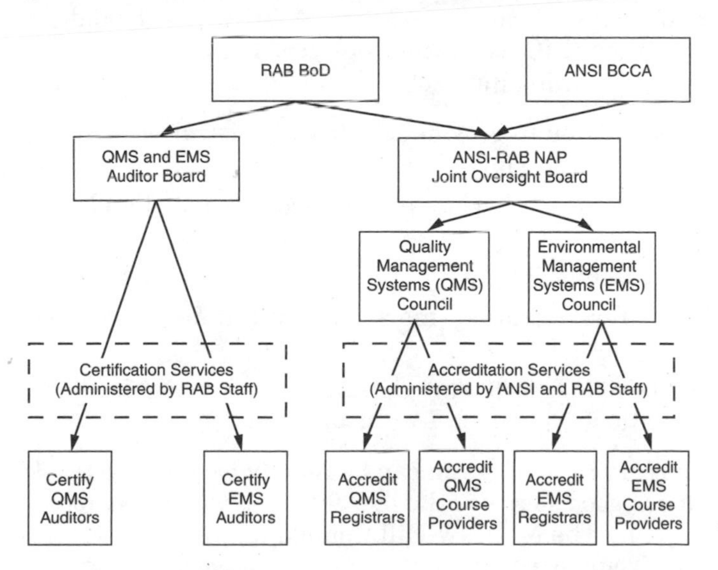

Figure 14-2. ANSI-RAB National Accreditation Program (NAP).

of this publication, 48 ISO 9000 registrars have been accredited by ANSI-RAB NAP.

ANSI-RAB NAP's primary mission is to provide assurance of the competence and reliability of registrars that audit and register organizations to the ISO 9000 and ISO 14000 standards. Its criteria for accrediting registrars is based on the following international standards and guides:

EN 45012	*General Criteria for Certification Bodies Operating Quality System Certification* (This standard is applied worldwide by all accreditation agencies to all registrars.)
ISO/IEC Guide 61	*General Requirements for Assessment and Accreditation of Certification / Registration Bodies*
ISO/IEC Guide 62	*General Requirements for Bodies Operating Assessment and Certification / Registration of Quality Systems*
ISO 10011	*Guidelines for Auditing Quality Systems* (to be replaced by ISO 19011, *Guidelines on Auditing Quality and Environmental Management Systems*)

ANSI-RAB NAP has signed memorandums of understanding (MOUs) with accreditation bodies in other countries to ensure mutual recognition of each other's accreditation systems. These bodies represent such countries as the United Kingdom, the Netherlands, Japan, Italy, Australia, and New Zealand.

While there are hundreds of registration bodies around the world, not all of them have internationally accepted accreditation. Therefore, you should carefully study the credentials of potential ISO 9000 registrars. First, make sure the registrar meets the appropriate North American Industry Classification System (NAICS) or Standard Industrial Classification (SIC) codes for your scope of business. Then ask the registrar the following questions:

- What body accredited you to provide ISO 9000 registration? (ANSI-RAB NAP, UKAS, RvA, or equivalent?)
- Do you have certified auditors who are qualified to conduct audits in my particular industry?
- Is your firm financially stable?
- Is your ISO 9000 registration mark recognized and accepted throughout the EU?
- Most especially: Is your ISO 9000 registration mark recognized and accepted in the nation(s) in which I want to do business?

Fortunately, such registrars exist and can be readily located if you ask the right questions. A good source is the *ISO 9000 Registered Company Directory,* published by The McGraw-Hill Companies.

The moral: Know your registrar!

Preliminary steps

The precise procedures for applying for registration vary among registrars. The sequence of events, however, is fairly common. Keep in mind that the process is greatly influenced by

- Facility size
- Facility scope
- Extent of documentation
- Participation of facility management
- Conformance of the quality system to the ISO standard

Smaller facilities with straightforward processes and by-the-numbers quality systems generally find the registration process to be a clean and brisk one. Larger facilities with complex operations and those with ill-defined or poorly documented quality systems can find the registration process to be extended and often quite difficult.

Sometimes, before initiating the registration process, registrars conduct a comprehensive preassessment of the facility. The purpose of this preassessment is to verify that the facility has at least the essentials of an ISO 9000 quality system in place. Facilities that have not had the assistance of experienced ISO 9000 auditors, consultants, and implementers would be wise to request such a preassessment as part of their ISO 9000 implementation process.

As Jim Ecklein of Augustine Medical says, "A preassessment helps your registrar to develop an understanding of your facility, process, and product. And it gives your people an idea of what audits are like. It calms their nerves. Once you've been through a preassessment, you won't be as nervous when it's time for the formal registration audit....My feeling is, if you don't mind failing your first registration audit, don't bother getting a preassessment."

The Registration Application

Normally, the first formal step is for the facility to file a formal application with the registrar. Some registrars charge a fee for this application, whereas others do not. Along with this application, the facility management usually completes a questionnaire about its operation. From the management responses, the registrar makes an initial judgment of the facility's state of readiness for ISO 9000 registration.

Most registrars also conduct an off-site audit of the facility's quality system documentation. This amounts to a review of the facility's quality manual or equivalent documentation. If the registrar determines, through its review of the questionnaire or the quality manual, that the quality system has serious nonconformances, it so informs the facility.

At this point, the registration process halts until the facility can show that it has corrected the nonconformances. Note that registrars are barred from providing guidance to correct nonconformances or give any further assistance, such as training and consulting services. The facility must obtain these services from other sources.

Timing and Costs

It is the goal of every consumer to get the best possible price for a good or service, whether it be a television, a car, or home furnishings, and ISO 9000 registration is no different. The two most frequently asked questions regarding registration are (1) How much will it cost? and (2) How long will it take?

As you enter the market for a registrar, you will find a wide price range for registration services. In determining fees, registrars typically look at

- Facility size
- Facility scope
- Extent of documentation
- Participation of facility management
- Quality system conformance to the ISO 9000 standard

Facility size tops the list of registration cost factors. On average, costs tend to be higher for larger facilities than for small or medium-sized companies. Organizations can spend several thousand dollars on registration once internal, external, and registrar costs are totaled.

Keeping costs to a minimum

There are three key elements that comprise the cost of registration:

- Daily rate
- Overhead expenses
- Travel and accommodations

Typically, most registrars charge a daily rate. While this charge is pretty straightforward, things can get confusing when overhead and travel expenses are added to it. Some registrars will quote a daily rate and then tack on extra charges for office preparation or other services. This creates confusion and presents an inaccurate picture of the total registration cost.

With travel, the cost is generally added on top of the registration fee. Therefore, it is a good idea to find out if the registrar has auditors located nearby or intends to fly them in from out of town.

To learn the total projected registration cost up front, you should obtain cost estimates for

- The registration application
- Document review
- Preassessment (optional)
- The registration audit
- Miscellaneous expenses, such as travel and lodging
- Surveillance audits (ISO 9000 generally requires one per year)

Ask the registrar to give you a quote on all fees expected to be incurred for the full 3-year registration cycle so that you can get an accurate estimate of the total cost. Be thorough and insistent, demanding a full accounting up front.

You should also be aware that while most ISO 9000–registered facilities have to seek reregistration after a 3 year period, some registrars offer a continuous surveillance option. Under this option, a facility's registration never expires as long as certain requirements are met, and surveillance audits, usually conducted every 6 months, reveal continuing conformance to the standard. This option can produce financial advantages for a company, because continuous surveillance eliminates the need for a full quality system audit every 3 years.

For larger companies, as a matter of keeping costs down, you may want to bring in company representatives from other locations to observe audit activities, especially during a preassessment. Your representatives can learn cost-cutting methods to apply to other facilities, making better use of internal resources in implementing the quality system.

While ISO 9000 registration may prove more expensive than originally estimated, consider the cost of not having a quality system. Since the ISO 9000 quality system standard is based on prevention, it is cost-effective because companies can avoid costly inspections, repeated service calls, and warranty costs while benefiting from reduced waste and fewer customer complaints and returns. Superior quality is critical for retaining current customers and helps companies to attract new customers.

It is a well-known fact that it costs much more—estimates range from 5 to 25 times more, depending on the industry—to attract a new customer than to retain a present one. No one can argue that the cost of poor quality is far greater than the cost of ensuring a superior quality system. Its savings include eliminating potential lost sales, the cost of redesign, corrective actions, unreported scrap, and extra manufacturing costs due to defects. As a result, many companies more than recoup the cost of ISO 9000 registration in annual cost savings.

While the cost of ISO 9000 registration can be somewhat significant, companies that want to be successful in the domestic and global markets cannot afford not to register, especially since registration is becoming the norm in industries around the world.

Timelines for registration

Like cost estimates, there is no set timeline for completing a registration audit. The number of days required varies, depending on several factors. Generally, the length of time required to complete a registration audit is determined by

- The size of the company
- The number of employees
- The complexity of a company's operations

The European Accreditation of Certification (EA), an organization of 17 European accreditation bodies, has issued EN 45012, *General Criteria for Certification Bodies Operating Quality System Certification,* the controlling standard for registrars. EN 45012 guidelines list the minimum days required for a valid registration audit (see Table 14-1). In evaluating your prospective registrar's proposed audit schedule, be sure to ask if it follows the EA's EN 45012 guidelines.

On average, it takes a company 6 months to a year to prepare for the registration audit, with less time needed if there is consultant assistance. The time required to register to ISO 9000 depends on the status of the facility's current quality system, its commitment to the process, and the resources it is willing to use.

A company that is just beginning the process and does not have a quality management system in place, or has one that is poorly documented, can expect a registration time frame of about 18 months if it writes all the documentation. If a consultant writes the documentation and assists in implementation, this time period could be reduced to as little as 6 months.

The registration audit process, from evaluation of documentation to issuance of a certificate, takes a minimum of 2 months to complete, assuming that there are no major problems with the quality system. It is important to remember that registering to ISO 9000 is a process. It is not something that can be done overnight. It requires patience, training, and the complete commitment of management.

Table 14-1. ISO 9000 Time Table Guidelines

	ISO 9001		ISO 9002	
No. of employees	A*	S†	A*	S†
1–4	2	1	2	1
5–9	2.5	1	2	1
10–14	3	1	2	1
15–19	3.5	1	3	1
20–29	4	1.5	3	1
30–59	6	2	5	2
60–99	7	2	6	2
100–249	8	2.5	6	2
250–499	10	8	8	2
500–999	12	4	10	3
1000–1999	15	5	12	4
2000–3999	18	6	14	5
4000–7999	21	7	17	6
8000+	24	8	19	6

Note: Surveillance audits are conducted once per year; or at months 9, 18, and 27 in a 3-year cycle. Figures are based on EA guidelines to EN 45012, Annex 2. Person days include documentation review, on-site audit, and preparation of the audit report.

*A=audit (person days).

†S=surveillance.

The Registration Audit

Once the registrar has satisfied itself that the quality system depicted in the quality manual conforms to ISO 9000 standards, it schedules an on-site audit of the facility. Usually, the client is allowed to pick the date for the registration audit. On that date, the organization should be prepared to undergo inspection of its facility, quality system, records, and other documentation by the audit team.

The audit team is composed of credentialed, accredited, and highly qualified quality system auditors. The standards for quality system auditor certification are set by bodies specially created for this purpose.

In the United States, quality system auditors are certified by the Registrar Accreditation Board *(RAB)*. The United Kingdom uses the International Register of Certificated Auditors *(IRCA)* for quality system auditor certification, under the aegis of the Department of Trade and Industry. The quality auditor certification body in Japan is the Japanese Registration of Certificated Auditors *(JRCA)*.

The International Audit and Training Certification Association (IATCA), a global quality system auditor certification and training course accreditation system, took effect in 1998. IATCA certification is primarily designed for auditors who frequently operate in more than one country and is intended to eliminate the need for multiple certifications.

This program for mutual recognition of auditor qualifications was started by IRCA in 1993. In 1994, an interim Multilateral Recognition Agreement (MLA) was signed by IRCA, RAB, the Joint Accreditation System–Australia New Zealand (JAS-ANZ), and the Quality Society of Australasia (QSA), under which each agreed to recognize certificates issued by the others.

IATCA was launched by a memorandum of understanding signed in 1995 by some of the world's largest certification bodies, representing 15 countries, with a total of 21 organizations eventually signing on. In addition to the four interim MLA signers, signatories included organizations representing Brazil, China, France, India, Japan, Malaysia, Singapore, South Africa, South Korea, and Taiwan. Each signatory agreed to develop certification programs according to mutually recognized criteria.

The signatories to the memorandum of understanding then adopted draft international criteria for auditor certification and training course accreditation. Participating organizations use the criteria as the basis for peer review of member programs. The IATCA scheme took effect in 1998 after at least three participating bodies successfully completed peer evaluations and signed MLAs for auditor certification and training course accreditation.

Initial signatories to the auditor certification MLA were IRCA, RAB, QSA, and the China National Registration Board for Auditors (CRBA). Signatories to the auditor training course accreditation MLA were IRCA, ANSI-RAB NAP, JAS-ANZ, and CRBA. Additional organizations will be added to the list of MLA signatories as they qualify through peer evaluations.

The qualifications and practices for quality system auditors are thoroughly documented:

- Under the old standards, ISO 10011-1, 10011-2, and 10011-3 spell out (1) guidelines for auditing quality systems, (2) qualification criteria for quality system auditors, and (3) management of audit programs, respectively. These guidelines apply to internal quality audits required by the standard, as well as second-party and third-party quality system audits. Under the 2000 revisions, ISO 10011 was to be merged into one document. Instead, it will be combined with the ISO 14010, 14011, and 14012 environmental auditing standards to form ISO 19011, *Guidelines on Auditing Quality and Environmental Management Systems.*
- ISO/IEC Guide 62 (1996) specifies guidelines for third-party auditing and registration of a supplier's quality system.

The audit team is made up of a *lead auditor,* who coordinates the audit and handles relations with the facility, and one or more other auditors, depending on facility size. Ideally, the team includes at least one auditor who is experienced in the type of operation being audited. Auditors must be free of any potential conflict of interest, and the facility has the right to object to the inclusion of any auditor with whom conflicts may arise.

Inspection of the facility

Armed with their copy of the quality manual, the audit team inspects the facility and its process. The purpose is to ensure that the quality system depicted in the manual

- Is adequate to the purposes of the process being audited
- Is capable of ensuring that output will conform to documented requirements
- Fairly represents the quality system being operated in the facility.

First, the audit team meets with facility management to review the audit procedure, set up lines of communication, and resolve any open issues. Then the on-site audit of the facility begins. This inspection entails

- A detailed review of the quality system to determine the extent to which it conforms to the requirements of the standard
- Facility tours, employee interviews, and reviews of documentation, including the written procedures and operating instructions referred to in the quality manual
- Checks of critical quality system functions such as internal audits, management reviews, corrective action, and documentation changes

In all respects, facility personnel are expected to demonstrate that the procedures documented in the quality manual are used in practice. The audit is detailed, specific, and rigorous. As you have no doubt observed, ISO 9000 is no façade. There is little wiggle room and no place to hide.

As Anthony Coggeshall of Adhesives Research observes, "Good auditors spend the bulk of their time talking not to management or supervisors, but to people who do the real work. Oh, they'll talk to all the big cheeses coming in and going out, but in between they talk to the troops. Makes it hard to hide things."

The audit report

At the end, the audit team makes a verbal report of its findings to facility management. The team may point out *observations,* problems of a relatively minor nature, as well as major and minor *nonconformances.* All major and minor nonconformances must be corrected within a designated timeframe before a registration certificate can be issued.

The audit team issues its findings in a report to the registrar. The registrar may conduct a full or partial reaudit to confirm that corrective actions have been taken. These reaudits can require the payment of additional fees.

When this phase is concluded, the registrar issues a certificate of registration.

Privileges of Registration

Henceforth the facility is registered to ISO 9001, 9002, or 9003. Under the 2000 revisions, it will be registered only to ISO 9001. As evidence, the facility is awarded a certificate that bears the registrar's mark, as well as the logo of the accrediting body (such as ANSI-RAB NAP or UKAS), if applicable. The certificate also bears a unique registration number.

Now the facility can show its certificate to customers and others as evidence of the objective approval of its quality system. It can use its registration logos in advertising and other printed matter, as regulated by its registrar's guidelines. Its name also may appear in published directories of registered firms, such as the *ISO 9000 Registered Company Directory,* published by The McGraw-Hill Companies, that are consulted by purchasing executives in the United States, the European Union, and elsewhere.

A facility also may find that ISO 9000 registration can greatly reduce the number of audits it faces. Ray Querciagrossa of Allmand Industries said, "As a result of our ISO 9002/QS-9000 certification, we now have a reduction in the amount of requests for completing customer surveys and audits. We're now a member of the club."

After Registration

Registration is not the end of the process. It is only the beginning. From that point forward, the facility and the registrar are, in effect, married.

The basic elements of the relationship are

- Payment of an annual fee
- Review and approval of quality system changes
- Regularly scheduled surveillance audits

Quality system changes

As the facility and process change, the quality system will require changes as well. Not only is change expected by ISO 9000, it is considered a virtual necessity.

The quality system's internal audit, management review, and corrective action mechanisms will generate continuous changes and improvements to specific quality system procedures. In the course of a surveillance audit, which amounts to a miniaudit of the facility, the registrar will evaluate these changes and their impact on the quality system's conformance to the standard.

Surveillance audits

The surveillance audit is performed to evaluate changes in the quality system, follow up on prior corrective actions, and verify that the quality system continues to conform to requirements. These audits are always performed, but are full audits only in unusual cases. Usually, they focus on

- Results of past corrective actions
- Areas that have generated nonconformances in the past
- Randomly selected elements of the quality system

The audit follows the same general routine as the registration audit, consisting of the

- Initial meeting
- Audit
- Verbal report of findings
- Written report of findings
- Corrective action requests, with deadlines

Most facilities find these audits to be helpful, not burdensome. "I enjoy the audits," says Anthony Coggeshall, "They are not looking to find things wrong with my system. They are looking to improve my company."

Ray Querciagrossa of Allmand Industries remarks, "The surveillance audits have worked very well for us. They force us to keep updated and never stop evolving our quality system."

Maintaining Registration

Registration is easy to keep, as long the bills are paid and the quality system remains consistent with the standard. However, registration can be suspended for serious nonconformances, misuse of registration logo or documents, and other serious rule violations. Suspension can last at least until the facility implements and documents corrective actions. Failure to correct serious nonconformances can result in termination of registration.

Barring these types of occurrences, most facilities find that ISO 9000 registration is easy to maintain. When fully implemented, ISO 9000 quality systems tend to become self-enforcing, self-reinforcing, and almost second nature, as central to the facility as the product or service it produces.

Registration Effects of the 2000 Revisions

Under transitional planning by the International Accreditation Forum (IAF), ISO Technical Committee (TC) 176, and the ISO Committee on Conformity Assessment (CASCO), registrars will not be able to issue ISO 9001: 2000 registration certificates before the revised standard is published. Prior to publication, registrars can conduct third-party audits to the latest draft of ISO 9001: 2000. Registration certificates issued to the 1994 ISO 9000 standards will remain valid for a maximum of 3 years after the revised standards are published.

Auditors and other relevant registrar personnel will be required to demonstrate knowledge and understanding of ISO 9001: 2000 requirements; the eight quality management principles on which the revised standards are based, which appear in ISO 9004: 2000, Element 4.3, and ISO 9000: 2000, Element 0.2, and the concepts and terminology of ISO 9000: 2000.

Registrars must take particular care in defining the scope of ISO 9001: 2000 registration certificates and the permissible exclusions from the standard's requirements, as set forth in ISO 9001:2000, Element 1.2. When issuing a tailored registration certificate, a registrar "may only exclude quality management system requirements that neither affect the organization's ability, nor absolve it from its responsibility, to provide product that meets customer and applicable regulatory requirements." These exclusions are limited to requirements within Element 7 (Product Realization) and may be due to the nature of the organization's product, customer requirements, or applicable regulatory requirements.

Chapter

15

The Future of ISO 9000

The pace at which ISO 9000 has developed is truly amazing. It was only in the late 1950s that MIL-Q-9858 was developed, the late 1960s that the British adopted DEF/STAN 05-8, the late 1970s that the British created BS 5750, and the late 1980s that ISO 9000 was first published.

Development and change will continue to be swift. As we have seen previously and will see again later in this chapter, ISO Technical Committee (TC) 176, Quality Management and Quality Assurance, which is responsible for the standard, is in the process of revising the standards to make them more user friendly. This total overhaul, which includes the merging of ISO 9002 and 9003 into ISO 9001, is to be finalized in November or December of 2000.

Meanwhile, it is likely that acceptance of ISO 9000 as the definitive international quality standard will continue to expand. Broader acceptance will happen even if the European Union (EU) never fully achieves total economic or political unity. In business, both national and multinational companies need a common quality language and a level quality playing field. If ISO 9000 did not exist, someone would have to invent it.

For all its shortcomings, and despite all the misunderstanding and misinformation that is routinely published, the ISO 9000 standard is unique. It transcends products and processes, it stands above language and custom, and it rests on principles of quality management that are almost universally accepted. In this chapter we will peer into our informed crystal ball to predict the future of ISO 9000.

Unification of the European Union

Since the emergence of the Single Market in 1993, the EU has become one of the world's largest trading powers. The EU is a major player on the global scene and a leading economic partner for most countries.

In the United States, the EU is a major partner. Some 40 percent of U.S. foreign investments can be attributed to the EU, making it one of the two top foreign markets for the United States. Conversely, the EU accounts for about 50 percent of foreign investments in the United States. Up to 3 million American jobs are due to EU investments.

Given the high level of acceptance that ISO 9000 has captured in the EU, it only makes sense for U.S. firms that have a significant market presence in the EU to obtain whatever quality system registration is required for domestic EU firms. ISO 9000 registration is especially important for facilities that make products already covered by EU product directives, for CE Marking product certification required by these directives cannot be obtained without quality system registration.

American facilities with future interests in Europe should consider obtaining ISO 9000 registration, irrespective of EU requirements. Even an American facility with no interest in the EU should consider using an ISO 9000 quality system model. It can improve quality, productivity, and customer satisfaction, which ultimately lead to a stronger competitive edge in the domestic or foreign marketplace.

Toward an American Accreditation Body

While the U.S. government has not officially sanctioned an American organization to accredit U.S.-based registrars, on a par with such organizations as the United Kingdom Accreditation Service (UKAS) in Great Britain and Raad voor Accreditatie (RvA) in the Netherlands, it has begun to recognize the importance of ISO 9000 to America's worldwide business interests. The U.S. Department of Defense (DoD) has canceled its long-time military purchasing quality standard MIL-Q-9858A, replacing it in new contracts with supplier requirements for ISO 9000 and other nongovernmental standards. Similarly, the National Aeronautics and Space Administration (NASA) and the General Services Administration (GSA) are promoting the use of ISO 9000 by their suppliers.

The U.S. Food and Drug Administration (FDA) has revised its Current Good Manufacturing Practices (CGMP) regulatory requirements to parallel ISO 13485 and 13488, harmonized versions of ISO 9001 and 9002 that contain specific requirements for medical device suppliers.

As previously mentioned, the American National Standards Institute (ANSI) and the Registrar Accreditation Board (RAB) created the National Accreditation Program (NAP) in 1996 to accredit ISO 9000 and ISO 14000 registrars. ANSI-RAB NAP has met all international standards and expectations and has achieved international acceptance and recognition.

The U.S. government's National Institute of Standards and Technology (NIST) has established the National Voluntary Conformity Assessment System Evaluation (NVCASE). NVCASE recognizes U.S. accreditation bodies in order to enable American manufacturers to satisfy conformity assessment requirements mandated by foreign governments. As of this writing, NVCASE had only recently begun operating and had yet to recognize any accreditation body. But ANSI-RAB NAP had been solicited to apply for recognition, and NVCASE officials expect the application to be approved.

Changes to the Standard: The 2000 Revisions

ISO 9000 is a living set of standards, always subject to change. International Organization for Standardization (ISO) protocols require all standards to be reviewed every 5 years to determine whether they should be confirmed, revised, or withdrawn.

The ISO 9000 standards were released in 1987. In 1990, ISO Technical Committee (TC) 176, Quality Management and Quality Assurance, adopted a two-phase revision process for the ISO 9000 standards. The first phase allowed limited changes to the standards and was completed in 1994. In 1996, ISO TC 176 reaffirmed the two-phase revision process, with the second phase to produce more thorough revisions.

To develop thoroughly revised standards, ISO TC 176 conducted a global user and customer needs survey in 1997. This survey found that customers and users wanted

- Revised standards that are more compatible with the ISO 14000 series of environmental management system standards
- Revised standards with a common structure based on a process model
- Provisions for tailoring ISO 9001 to omit inapplicable requirements
- ISO 9001 requirements to include demonstration of continuous improvement and prevention of nonconformities
- ISO 9001 to address effectiveness, with ISO 9004 addressing both efficiency and effectiveness
- ISO 9004 to help achieve benefits for all interested parties, such as customers, owners, employees, suppliers, and society

- Revised standards that are simple to use, easy to understand, and contain clear language and terminology
- Revised standards that facilitate self-evaluation
- Revised standards that are suitable for all sizes of organizations, operating in any economic or industrial sector, with the manufacturing orientation of the old standards removed

The First Committee Draft (CD1) of these thoroughly revised standards was released by ISO TC 176 Subcommittee (SC) 2, Quality Systems, which is responsible for maintaining the ISO 9000 standards, in July 1998. It was followed in February 1999 by the Second Committee Draft (CD2) and in November 1999 by the Draft International Standard (DIS), the version of the 2000 revisions that is cited throughout this book. DIS will be followed in the third quarter of 2000 by the Final Draft International Standard (FDIS), with the new standards, including a single ISO 9001 registration standard, to be published in November or December 2000.

The move is a major part of the continuing series of steps planned to keep the ISO 9000 standard series dynamic and responsive to marketplace needs while making it more user friendly, according to former ISO TC 176 Chairman Reginald Shaughnessy. "We want a standard that is valid and acceptable to all industry sectors, that doesn't add unnecessary costs, that doesn't restrict trade, and that avoids a proliferation of standards," he said.

Currently, there are more than 20 ISO 9000 standards and documents, and this proliferation has raised concerns among ISO 9000 users and consumers. In response, ISO TC 176 agreed that the 2000 revisions will consist of four primary standards, supported by several technical reports. The four primary standards are

- ISO 9000: *Quality Management Systems—Fundamentals and vocabulary.* This document will simplify the quality language, which consists of the terms, definitions, and principles that accompany the standard. It is expected to clear up confusion about the standard's meaning by combining the terms and definitions in ISO 8402 and 9000-1 into one easy-to-understand document. In addition to a revised vocabulary, this standard will include an introduction to quality concepts.
- ISO 9001: *Quality Management Systems—Requirements.* As noted earlier, this document replaces ISO 9001, 9002, and 9003. It addresses an organization's quality management requirements in order to demonstrate its capability to meet customer requirements and applies to all generic product categories, such as hardware, soft-

ware, processed materials, and services. This standard can be tailored to fit an organization's operations through reduction in scope, thereby eliminating the need for ISO 9002 and 9003. Element 1.2 states, "The organization may only exclude quality management system requirements that neither affect the organization's ability, nor absolve it from its responsibility, to provide product that meets customer and applicable regulatory requirements. These exclusions are limited to those requirements within Clause 7 (Product Realization) (see also 5.5.5), and may be due to the following: (a) the nature of the organization's product; (b) customer requirements; (c) the applicable regulatory requirements. Where permissible exclusions are exceeded, conformity to this International Standard should not be claimed. This includes situations where the fulfillment of regulatory requirements permits exclusions that exceed those allowed by this International Standard."

- ISO 9004: *Quality Management Systems—Guidelines for performance improvement.* This standard, which replaces ISO 9004-1, provides guidance beyond ISO 9001 toward developing a comprehensive quality management system to improve an organization's overall performance. While following the same structure as ISO 9001: 2000, it will not be an ISO 9001 implementation or conformance guide. Instead, it gives guidance on all aspects of a quality management system, based on the principles of customer focus, leadership, involvement of people, systems approach to management, continual improvement, factual approach to decision making, and mutually beneficial supplier relationships. ISO 9004 will be useful in auditing the effectiveness of a company's quality management system, with the goal of achieving benefits for all stakeholder groups through sustained customer satisfaction.
- ISO 19011: *Guidelines on Auditing Quality and Environmental Management Systems.* Initially, ISO TC 176 was to draft ISO 10011: *Guidelines for Auditing Quality Systems,* which would integrate and replace the quality system auditing requirements of ISO 10011-1, 10011-2, and 10011-3. Instead, the Joint Working Group (JWG) on Quality and Environmental Auditing, consisting of experts from ISO TC 176, Quality Management and Quality Assurance, SC 3, Auditing, and ISO TC 207, Environmental Management, SC 2, Environmental Auditing, agreed to draft a joint quality and environmental management systems auditing standard, ISO 19011. It will apply to first-, second-, and third-party audits; auditor qualification criteria; and audit program management. It is expected to be released in early 2001, replacing ISO 10011-1, 10011-2, and 10011-3, along with environmental auditing standards ISO 14010, 14011, and 14012.

As of this writing, the remaining old ISO 9000 standards are being reviewed by ISO TC 176 for incorporation within the four revised standards, withdrawal, or reissue as technical reports.

The revised standards use a simple process-based structure that is more generic than the old 20-element structure, is consistent with the plan-do-check improvement cycle used in the ISO 14000 environmental management systems standards, and adopts the process management structure widely used in business today.

As mentioned in Chap. 12, the major clauses in the revised standards are

- Management responsibility (policy, objectives, planning, quality management system, management review)
- Resource management (human resources, information, facilities)
- Product realization (customer satisfaction, design, purchasing, production, calibration)
- Measurement, analysis, and improvement (audit, process and product control, improvement)

The ISO 9001 revision will not require an organization to rewrite its quality system documentation. In providing for a smooth transition to the revised standards, there will be no quality system documentation layout or structural requirements. Documentation should continue to reflect the organization's way of doing business.

Many of the revisions in structure, content, language, and terminology are designed to achieve greater compatibility with the ISO 14001 environmental management systems standard. ISO TC 176 is coordinating development of the revised standards with the work of ISO TC 207, Environmental Management, SC 1, Environmental Management Systems, to ensure that the two series of standards can be jointly implemented while avoiding duplication and conflicts. One result of this coordination is development of the previously mentioned ISO 19011 joint auditing standard.

Voices of the Users: The Future of ISO 9000

Brian Burke, Container Products Inc.:

> I'm sold on ISO 9000....I think that the concept of unifying all these facets of quality management and certifying the system by an objective third party is a good way to go. For us, the discipline of going through the preparation has been the most valuable thing. The registration will be neat, our commercial people will run with it, but the discipline of putting together the quality system and the documentation has been the most beneficial thing of all.

Dennis Beckley, Dayton-Rogers Corp.:

Our company has 3000 customers. This plant alone has 600. I probably undergo 20 mail-in audits and two on-site audits each month. I know ISO 9000 won't reduce my audit load in the near future, but in time it will. I know I'll be looking for ISO registration among my supply base, just as I look for Q1, and when my suppliers get ISO registered, I'll waive the audits.

Hopefully, as time goes by ISO 9000 will be recognized as a basic quality criterion by everyone: government, automotive, everyone. Obviously everyone has individual specs, but I think there needs to be a general quality standard that everyone adopts across the board.

Linda Kabel, Menasha Corp.:

I just hope ISO 9000 isn't just a passing fad. So many programs are here today, gone tomorrow. We're putting money and time into ISO 9000 because we think it's a good system, very in depth. Hopefully, in time, our customers will exempt us from audits, once we're registered. That would be a real time saver.

David Turteltaub, Phillips Circuit Assembly:

I think the ISO 9000 standard itself is good. I don't see much need for improvements or changes to the standards themselves. I do think the expense has to be looked at. There are smaller companies who can't afford it.

But the United States is still number one exporter in the world. I think we need to get hold of ISO 9000 and run with it, instead of just letting the EU dictate it to us. I think if we take the lead on ISO 9000, it'll take hold worldwide. It will be around for a long time.

Don Van Hook, Strahman Valves:

For one thing, I think there should be international accreditation, rather than separate accreditation for each country. I also think ISO might want to get just a little more specific in some areas and not leave so much open to interpretation. Record retention, for example. Our consultant recommended 10 years. Why? Because it was a nice round number. We suggested three to five years, and he agreed that was probably adequate.

George Raub, TRW:

I'm comfortable with specifics of the standard itself. It covers the minimum kind of quality system. It ought to go more in depth on quality costs and continuous improvement, but a well-run company looks at those things, whether a standard requires them to or not.

The registration process needs to be cleaner. Our government needs to get treaties for cross recognition of accreditation and certification. It's a shame that our government makes it so confusing for U.S. businesses to pursue what is, after all, in the national interest: building American exports.

My fearless forecast is that the next hotbed of ISO acceptance will be the Far East. Not Japan, which is already inbred with quality circles and work groups and the like, but the Third World: India, Asia and the like.

Anthony Coggeshall, Adhesives Research, Inc.:

Looking into the future, within two years, give or take a year, you won't be able to do business in Europe without ISO 9000 registration. Within five years, you won't be able to work with a U.S. government agency, or a company supplying a government agency, without ISO 9000 certification. Within five to 10 years, it'll be tough to do business in the United States without it.

Part

4

ISO 9000 Derivatives

Chapter

16

Auto Industry Embraces ISO 9000

Perhaps more than any other business sector, the automobile industry has accepted ISO 9000, approved its guidelines, and created its own sector-specific ISO 9000 derivatives. On a national basis, these derivatives were designed to harmonize individual automakers' quality systems, thereby eliminating conflicting requirements for suppliers.

In the United States, the Big 3, Chrysler (now DaimlerChrysler), Ford, and General Motors, developed QS-9000 and the TE Supplement. In Europe, automakers and their major suppliers brought forth VDA 6.1 in Germany, EAQF in France, and AVSQ in Italy. For international suppliers, these multiple ISO 9000 automotive derivatives, with sometimes conflicting and overlapping requirements, have in turn led to a degree of confusion and duplication.

Further harmonization was needed, and an international automotive quality standard, ISO/TS 16949 was developed. As of this writing, ISO/TS 16949 is available as an alternative to the national automotive standards. However, this arrangement may prove to be temporary. Once the 2000 revisions to ISO 9000 take effect, ISO/TS 16949 will itself be revised to incorporate these changes. After this process is completed, ISO/TS 16949 may replace all the national automotive quality standards. This chapter examines these automotive derivatives of ISO 9000.

QS-9000

Before the development of Quality System Requirements QS-9000, each of the automotive Big 3, along with truck manufacturers, had its own proprietary quality standard. These standards included the Ford Q-101 Quality System Standard, the General Motors North American Operations (NAO) Targets for Excellence (TFE), and the Chrysler *Supplier Quality Assurance* manual.

These standards were based on a variety of definitions and measurements of quality and frequently proved to be contradictory. These contradictions multiplied as the standards evolved, becoming more demanding and more specific in requiring suppliers to use certain types of quality tools, techniques, and reporting systems. As a result, some suppliers were subjected to as many as 60 to 70 audits a year.

In 1988, the Big 3, along with truck manufacturers, realized that subscribing to one commonly used quality standard would make doing business with suppliers easier and more efficient. Forming the Chrysler/Ford/General Motors Supplier Quality Requirements Task Force, they began to develop a harmonized automotive industry standard, using ISO 9000 as the framework.

ISO 9000 appealed to the automakers because it was already working in other industries and met most of the existing Big 3 quality program requirements. QS-9000 development took 6 years, with the first edition released in 1994. It was followed by the second edition, which contained minor revisions, in 1995 and the substantially revised third edition in 1998.

QS-9000 applies to all internal and external suppliers of production or service parts; production materials; and heat treating, painting, plating, or other finishing services directly relating to the Big 3 or other OEM (original equipment manufacturer) customers.

Section I of QS-9000 contains all 20 elements of ISO 9001, with most incorporating additional automotive sector–specific requirements. Not all quality requirements could be harmonized, however. As a result, Section II of QS-9000 consists of customer-specific requirements, which are additional elements addressing the individual product needs of DaimlerChrysler, Ford, General Motors, and six truck manufacturers, namely, Mack Trucks, Inc., Navistar International Transportation Corp., PACCAR, Inc., Volvo Truck North America, Mitsubishi Motors-Australia, and Toyota Australia.

DaimlerChrysler and General Motors require their production and service part suppliers to register to QS-9000. As with ISO 9002, Element 4.4 does not apply to suppliers that are not responsible for design.

Sector-specific guidelines

Let's look at the automotive sector–specific guidelines that the Big 3 has sprinkled throughout the ISO 9000–based requirements in Section I of QS-9000. Nearly every ISO 9001 element contains some of these additional requirements.

- *4.1.2.4 Organizational Interfaces.* Suppliers are required to implement a system to inform management of appropriate activities during concept development, prototype, and production, following guidelines in the *Advanced Product Quality Planning and Control Plan* (APQP) reference manual. Suppliers must use a multidisciplinary approach to decision making and must have the ability to communicate necessary information and data in the customer-prescribed format.

- *4.1.3.1 Management Review.* Management review must include all elements of the entire quality system.
- *4.1.4 Business Plan.* Suppliers are required to use formal, documented, and comprehensive business plans. The business plan is a controlled document whose content is not subject to third-party audit. Business plans should cover short-term (1 to 2 years) and long-term (3 years or more) goals and should be based on analysis of competitive products, as well as benchmarking inside and outside the automotive industry. These plans must adhere to current and future customer expectations and include an objective process for collecting information. Suppliers must document methods to ensure that plans can be easily tracked, updated, revised, and reviewed.
- *4.1.5 Analysis and Use of Company-Level Data.* Suppliers must document their trends in quality, operational performance, productivity, efficiency, effectiveness, and current quality levels for key product and service features. These trends should be compared with those of competitors, appropriate benchmarks, and progress toward overall business objectives. This information should be used to develop priorities to solve customer-related problems and support customer-related status review, decision making, and longer-term planning.
- *4.1.6 Customer Satisfaction.* Suppliers must have a documented process to determine customer satisfaction, including frequency of determination and how objectivity and validity are ensured. Trends in customer satisfaction and key indicators of customer dissatisfaction should be documented and supported by objective information. These trends should be compared with those of competitors or other relevant benchmarks and reviewed by senior management.
- *4.2.3.1 Advanced Product Quality Planning.* Suppliers must establish and implement an advanced product quality planning process. Internal multidisciplinary teams should be established to prepare plans for the production of new or altered products. These teams should use techniques outlined in the APQP reference manual and are responsible for developing and finalizing special characteristics, developing and reviewing failure mode effects analyses (FMEAs) and control plans, and establishing actions to reduce the potential failure modes.
- *4.2.3.2 Special Characteristics.* When special characteristics are identified on the customer design record, the supplier's process control

guidelines and similar documents, such as FMEAs, control plans, and operator instructions, must be marked with the customer's special characteristic symbol or the supplier's equivalent symbol or notation.

- *4.2.3.3 Feasibility Reviews.* Before a supplier agrees to enter into a contract to manufacture a product, it must first investigate and confirm the product's manufacturing feasibility. The supplier must determine if the design, the material, and the process planned to produce the product are suitable and whether the proposed product conforms to all engineering requirements at the required statistical process capability and at specified volumes.
- *4.2.3.4 Product Safety.* Due care and product safety must be considered in the supplier's design control and process control policies and practices. The supplier should promote internal awareness of safety conditions relative to its product.
- *4.2.3.5 Process Failure Mode and Effects Analyses (Process FMEAs).* Process FMEAs must consider all special characteristics. Furthermore, suppliers must improve the FMEA process to emphasize defect prevention rather than defect detection.
- *4.2.3.6 Mistake-Proofing.* A mistake-proofing process must be in place to prevent the manufacture of nonconforming product. When the sources causing the nonconformities are identified, by FMEAs, capability studies, and service reports, the sources must be addressed by using the mistake-proofing methodology. Mistake-proofing must be used during the planning of processes, facilities, equipment, and tooling, as well as during problem resolution.
- *4.2.3.7 The Control Plan.* Suppliers must develop control plans at the system, subsystem, component, and/or material levels as appropriate. Control plans are written descriptions of the systems that control parts and processes. They address the important characteristics and engineering requirements of the product. The control plan can encompass processes for producing bulk materials, such as steel, plastic, resin, and paint, as well as those producing parts. Control plans must be revised or updated when products or processes differ significantly from those in current production. The control plan should list the controls used for process control and must cover the prototype, prelaunch, and production phases, as appropriate.
- *4.2.4.1 Product Approval Process—General.* The supplier must fully comply with all requirements set forth in the *Production Part Approval Process* (PPAP) reference manual. PPAP applies to all production and service part manufacturing and changes. The customer must inform the supplier of appropriate documents and items. Suppliers must retain a complete record for all part submissions. Parts cannot be shipped without customer approval.

- *4.2.4.2 Subcontractor Requirements.* Suppliers must use a part approval process, such as PPAP, for subcontractors.
- *4.2.4.3 Engineering Change Validation.* The supplier must verify that engineering changes are properly validated.
- *4.2.5.1 Continuous Improvement—General.* The supplier must continuously improve in quality, service (including timing and delivery), and price to benefit all customers. Continuous improvement extends to product characteristics, with the highest priority on special characteristics. Suppliers should develop a prioritized action plan for continuous improvement in processes that have demonstrated stability, acceptable capability, and performance.
- *4.2.5.2 Quality and Productivity Improvements.* The supplier must identify opportunities for quality and productivity improvement and implement appropriate improvement projects.
- *4.2.5.3 Techniques for Continuous Improvement.* The supplier must demonstrate knowledge of appropriate continuous improvement measures and methodologies and must use those which are appropriate.
- *4.2.6.1 Facilities, Equipment, and Process Planning Effectiveness* The supplier must use a multidisciplinary approach to develop facilities, processes, and equipment plans in conjunction with the advanced quality planning process. Plant layouts should minimize material travel and handling, facilitate synchronous material flow, and maximize value-added use of floor space. Methods must be developed to evaluate the effectiveness of existing operations and processes.
- *4.2.6.2 Tooling Management.* The supplier must establish and implement a system for tooling management. It must provide appropriate technical resources for tool and gauge design, fabrication, and full dimensional inspection. If any of this work is subcontracted, a tracking and follow-up system must be established.
- *4.4.1.1 Use of Design Data.* Suppliers must have a process to deploy information gained from previous design projects to current and future projects of a similar nature.
- *4.4.2.1 Required Skills.* The supplier's design activity should be qualified in such appropriate skills as geometric dimensioning and tolerancing (GD&T), quality function deployment (QFD), design for manufacturing (DFM)/design for assembly (DFA), value engineering (VE), design of experiments (DOE), failure mode and effects analysis (DFMEA/PFMEA, etc.), finite-element analysis (FEA), solid modeling, simulation techniques, computer-aided design (CAD)/computer-aided engineering (CAE), and reliability engineering plans.
- *4.4.4.1 Design Input—Supplemental.* The supplier must have appropriate resources and facilities to use computer-aided product

design, engineering, and analysis. If these functions are subcontracted, the supplier must provide technical leadership.

- *4.4.5.1 Design Output—Supplemental.* The supplier's design must result from a process that includes efforts to simplify, optimize, innovate, and reduce waste, such as QFD, VE, DFM/DFA, DOE, tolerance studies, response surface methodologies, or appropriate alternatives; use of GD&T as applicable; analysis of cost, performance, and risk tradeoffs; use of feedback from testing, production, and the field; and use of design FMEAs.
- *4.4.8.1 Design Validation—Supplemental.* Design validation must be performed in conjunction with customer program timing requirements. Validation results must be recorded, and design failures must be documented in the validation records. Procedures for corrective and preventive actions must be followed in addressing such design failures.
- *4.4.9.1 Design Changes—Supplemental.* All design changes, including those proposed by subcontractors, must have written customer approval, or waiver of such approval, before the supplier can begin production. For proprietary designs, impact on form, fit, function, performance, and/or durability must be determined with the customer so that all effects can be properly evaluated.
- *4.4.9.2 Design Change Impact.* The supplier must consider the impact of a design change on the system in which the product is used.
- *4.4.10 Customer Prototype Support.* When required by the customer, the supplier must have a comprehensive prototype program using the same subcontractors, tooling, and processes as are used in production wherever possible. Performance tests must consider and include as appropriate product life, reliability, and durability. All performance testing activities must be tracked to monitor timely completion and conformance to requirements. When these services are contracted, the supplier must provide technical leadership.
- *4.4.11 Confidentiality.* The supplier must ensure the confidentiality of customer-contracted products under development and related product information.
- *4.5.2.1 Engineering Specifications.* The supplier must establish a procedure to ensure the timely review, in business days, not weeks or months, along with distribution and implementation of all customer engineering standards, specifications, and changes. The supplier must maintain a record of the date on which each change was implemented in production. Implementation must include updates to all appropriate documents.

- *4.6.1.1 Approved Materials for Ongoing Production.* If the customer has an approved subcontractor list, then the supplier must purchase relevant materials from the subcontractors on that list. Any additional subcontractors may only be used after they have been added to list by the customer's materials engineering activity.
- *4.6.1.2 Government, Safety, and Environmental Regulations.* All purchased materials used in parts manufacturing must satisfy current government and safety constraints on restricted, toxic, and hazardous materials, as well as environmental, electrical, and electromagnetic considerations applicable to the country of manufacture and sale.
- *4.6.2.1 Subcontractor Development.* The supplier must carry out subcontractor quality system development with the goal of subcontractor compliance to QS-9000. Audits, if part of subcontractor development, should occur at supplier-specified frequency. Subcontractor audits to QS-9000 by the original equipment manufacturer (OEM) customer, an OEM customer-approved second party, or an accredited registrar will be recognized in lieu of audits by the supplier. The use of customer-designated subcontractors does not release the supplier from the responsibility for ensuring the quality of subcontracted parts, materials, and services.
- *4.6.2.2 Scheduling Subcontractors.* The supplier must require 100 percent on-time delivery performance from subcontractors. The supplier must supply appropriate planning information and purchase commitments to enable subcontractors to meet this expectation. The supplier must implement a system to monitor subcontractor delivery performance, with corrective actions taken as appropriate. Records of premium freight must include both supplier- and subcontractor-paid charges.
- *4.7.1 Customer-Owned Tooling.* Customer-owned tools and equipment must be permanently marked so that the ownership of each item is visually apparent
- *4.9.b.1 Cleanliness of Premises.* The supplier must maintain premises in a state of order, cleanliness, and repair appropriate to the product(s) manufactured.
- *4.9.b.2 Contingency Plans.* The supplier must prepare contingency plans, such as for utility interruptions, labor shortages, or key equipment failure, to reasonably protect the customer's supply of product in the event of emergency, excluding natural disaster and acts of God.
- *4.9.d.1 Designation of Special Characteristics.* The supplier must comply with all customer requirements for designation, documenta-

tion, and control of special characteristics. The supplier must provide documentation showing compliance with these customer requirements, as requested.

- *4.9.g.1 Preventive Maintenance.* The supplier must identify key process equipment and provide appropriate resources for machine and equipment maintenance and develop an effective planned total preventive maintenance system.
- *4.9.1 Process Monitoring and Operator Instructions.* The supplier must prepare documented process monitoring and operator instructions for all employees having responsibilities for operating processes. These instructions must be accessible at workstations and must be derived from the sources listed in the APQP reference manual.
- *4.9.2 Maintaining Process Control.* The supplier must maintain or exceed process capability or performance as approved via PPAP. To accomplish this, the supplier must ensure that the control plan and process flow diagram are implemented, including, but not limited to, adherence to the specified measurement technique, sampling plans, acceptance criteria, and reaction plans when acceptance criteria are not met. Significant process events, such as tool change or machine repair, should be noted on the control charts. When process and/or product data indicate a high degree of capability, the supplier may revise the control plan, as appropriate. The supplier must initiate the appropriate reaction plan from the control plan for characteristics that are identified on the control plan and are either unstable or noncapable. A supplier corrective action plan must then be completed, indicating specific timing and assigned responsibilities to ensure that the process becomes stable and capable. The plans are to be reviewed with and approved by the customer when so required.
- *4.9.3 Modified Process Control Requirements.* In the event that a customer sets higher or lower capability or performance requirements, these changes must be reflected in the control plan.
- *4.9.4 Verification of Job Setups.* Job setups must be verified whenever performed, such as initial run of a job, material changeover, job change, or significant time periods lapsed between runs. Job instructions must be available for setup personnel. Last-off part comparisons are recommended. The supplier must use statistical verification where applicable.
- *4.9.5 Process Changes.* The supplier must maintain records of process change effective dates.
- *4.9.6 Appearance Items.* For suppliers manufacturing parts designated by the customer as appearance items, the supplier must provide appropriate lighting for evaluation areas; masters for color,

grain, gloss, metallic brilliance, texture, and directness of image, as appropriate; maintenance and control of appearance masters and evaluation equipment; and verification that personnel making appearance evaluations are qualified to do so.

- *4.10.1.1 Acceptance Criteria for Attribute Characteristics.* Acceptance criteria for attribute data sampling plans shall be zero defects. Appropriate acceptance criteria for all other situations, such as visual standards, must be documented by the supplier and approved by the customer.
- *4.10.2.4 Incoming Product Quality.* The supplier's incoming quality system must use receipt and evaluation of statistical data by the supplier; receiving inspection and/or testing, such as sampling based on performance; or second- or third-party audits of subcontractor sites, when coupled with records of acceptable quality performance; or part evaluation by accredited laboratories.
- *4.10.4.1 Layout Inspection and Functional Testing.* A layout inspection and functional verification to applicable customer engineering material and performance standards must be performed for all products at a frequency established by the customer. Results must be available for customer review.
- *4.10.4.2 Final Product Audit.* The supplier must conduct audits of packaged final product to verify conformance to specified requirements, such as product, packaging, and labeling, at an appropriate frequency.
- *4.10.6.1 Laboratory Quality Systems.* The laboratory (supplier's testing facility, such as chemical, metallurgical, reliability, test validation, and fastener labs) must have a laboratory scope. The laboratory must document all its policies, systems, programs, procedures, instructions, and findings that enable it to ensure the quality of the tests or calibration results it generates within the scope.
- *4.10.6.2 Laboratory Personnel.* The personnel making testing and/or calibration professional judgments must have the appropriate background and experience.
- *4.10.6.3 Laboratory Product Identification and Testing.* The laboratory must have procedures for the receipt, identification, handling, protection, and retention or disposal of test samples and/or calibration equipment items, including all provisions necessary to protect the integrity of the items. The items must be retained until final data are complete throughout the life of the item in the laboratory, enabling traceability from final data to raw data.
- *4.10.6.4 Laboratory Process Control.* The laboratory must monitor, control, and record environmental conditions as required by relevant

specifications or where they may influence the quality of results. Requirements for environmental conditions, such as biologic sterility, dust, electromagnetic interference, radiation, humidity, electrical supply, temperature, and sound and vibration levels, must be established and maintained as appropriate to the technical activities concerned.

- *4.10.6.5 Laboratory Testing and Calibration Methods.* The laboratory must use test and/or calibration methods, including those for sampling, that meet the needs of the customer and are appropriate for the tests and/or calibrations it undertakes, preferably the current issue of those published as international, regional, or national standards. The laboratory must verify its capability to perform to the standard specifications before carrying out such work. When it is necessary to employ methods not covered by standard specifications, these are subject to agreement with the customer.
- *4.10.6.6 Laboratory Statistical Methods.* Appropriate statistical techniques should be applied to verification activities whose deliverables are data.
- *4.10.7 Accredited Laboratories.* Commercial or independent laboratories used by the supplier must be accredited. An accredited laboratory is one that has been reviewed and approved by an accreditation body such as the American Association for Laboratory Accreditation or the Standards Council of Canada.
- *4.11.2.b.1 Calibration Services.* Calibration of inspection, measuring, or test equipment must be conducted by a qualified in-house laboratory, a qualified commercial or independent laboratory, or a customer-recognized government agency. The laboratory scope must include the calibration of such equipment. Commercial or independent calibration facilities must be accredited to ISO/IEC 17025 or national equivalent or have evidence, such as an OEM customer or an OEM customer-approved second-party audit, that they meet the intent of ISO/IEC 17025 or national equivalent.
- *4.11.3 Inspection, Measuring, and Test Equipment Records.* Calibration activity records for all gauges, and measuring and test equipment, including those owned by employees, must include revisions following engineering changes (if appropriate); any out-of-specification readings as received for calibration; statements of conformance to specification after calibration; and notification to the customer if suspect material or product may have been shipped.
- *4.11.4 Measurement Systems Analysis.* Appropriate statistical studies must be conducted to analyze the variation present in the results of each type of measuring and test equipment system. This requirement applies to measurement systems referenced in the control

plan. The analytical methods and acceptance criteria used should conform to those in the *Measurement Systems Analysis* (MSA) reference manual, such as bias, linearity, stability, repeatability, and reproducibility studies. Other analytical methods and acceptance criteria may be used if approved by the customer.

- *4.12.1 Supplemental Verification.* When required by the customer, additional verification or identification requirements must be met.
- *4.13.1.1 Suspect Product.* This element applies to suspect material or product, as well as nonconforming product.
- *4.13.1.2 Visual Identification.* The supplier must provide visual identification of any nonconforming or suspect material or product and any quarantine areas.
- *4.13.2.1 Prioritized Reduction Plans.* The supplier must quantify and analyze nonconforming product and establish a prioritized reduction plan. Progress made toward the plan should be tracked.
- *4.13.3 Control of Reworked Product.* Rework instructions must be accessible and used by the supplier's appropriate personnel in their work areas. No rework may be visible on the exterior of products supplied for service applications without prior approval from the customer's service parts organization.
- *4.13.4 Engineering Approved Product Authorization.* The supplier must obtain prior customer authorization whenever a product or process is different from that currently approved. This applies equally to products or services purchased from subcontractors. The supplier must concur with any subcontractor requests before submission to the customer. The supplier must keep a record of the expiration date or quantity authorized. The supplier also must ensure compliance with the original or superseding specifications and requirements when the authorization expires. Material shipped on an authorization must be identified properly on each shipping container.
- *4.14.1.1 Problem-Solving Methods.* The supplier must use disciplined problem-solving methods when an internal or external nonconformance to specifications or requirements occurs. When external nonconformances occur, the supplier must respond in a manner prescribed by the customer.
- *4.14.1.2 Mistake-Proofing.* The supplier must use mistake-proofing methodology in its corrective and preventive actions process to a degree appropriate to the magnitude of the problems and commensurate with the risks encountered.
- *4.14.2.1 Returned Product Test/Analysis.* The supplier must analyze parts returned from customers' manufacturing plants, engineering

facilities, and dealerships. Records of these analyses must be kept and made available on request. The supplier must perform effective analysis and, where appropriate, initiate corrective action and process changes to prevent recurrence.

- *4.14.2.2 Corrective Action Impact.* Where applicable, the supplier must apply the corrective action taken and the controls implemented to eliminate the cause of a nonconformity to similar processes and products.
- *4.15.3.1 Inventory.* The supplier must use an inventory system to optimize inventory turns over time, ensure stock rotation, and minimize inventory levels.
- *4.15.4.1 Customer Packaging Standards.* The supplier must comply with all unique customer packaging standards or guidelines, including applicable service part packaging standards.
- *4.15.4.2 Labeling.* The supplier must develop a system to ensure that all materials shipped are labeled according to customer specifications.
- *4.15.6.1 Supplier Delivery Performance Monitoring.* The supplier must establish systems to support 100 percent on-time shipments to meet customer production and service requirements. If this goal is not maintained, the supplier must implement corrective action to improve delivery performance, including communication of delivery problem information to the customer. The supplier must have a systematic approach to develop, evaluate, and monitor adherence to established lead-time requirements. The supplier must implement a system to monitor performance to customer delivery requirements, with corrective action taken as appropriate. Records of supplier-responsible premium freight must be maintained. The supplier must ship all materials in conformance with customer requirements, adhering to up-to-date customer-specified transportation mode, routings, and containers.
- *4.15.6.2 Production Scheduling.* The supplier's production scheduling activity must be order-driven.
- *4.15.6.3 Electronic Communication.* The supplier must have a computerized system for receipt of customer planning information and ship schedules, unless waived by the customer.
- *4.15.6.4 Shipment Notification System.* The supplier must have a computerized system for on-line transmittal of advance shipment notifications (ASNs), transmitted at the time of shipment, unless waived by the customer. The supplier must have a backup method in case the on-line system fails. The supplier must verify that all ASNs match shipping documents and labels.

- *4.16.1 Record Retention.* Production part approvals, tooling records, purchase orders, and amendments must be maintained for the length of time the part or family of parts is used for production and service requirements, plus 1 calendar year, unless otherwise specified by the customer. Quality performance records, such as control charts and inspection and test results, must be retained for 1 calendar year beyond the year they were created. Records of internal quality system audits and management review must be retained for 3 years. Retention periods longer than those just specified may be specified by a supplier in its procedures. The supplier must eventually dispose of records. This requirement does not supersede any government requirements. All retention periods are considered minimums.
- *4.17.1 Internal Audit Schedules.* Internal auditing should cover all shifts and be conducted according to an annually updated audit schedule. When internal or external nonconformances or customer complaints occur, the planned audit frequency should be increased.
- *4.18.1 Training Effectiveness.* Training effectiveness must be reviewed periodically.
- *4.19.1 Feedback of Information from Service.* The supplier must establish and maintain a procedure for communicating information on service concerns to manufacturing, engineering, and design activities.
- *4.20.3 Selection of Statistical Tools.* Statistical tools, if applicable, for each process should be determined during advanced quality planning and must be included in the control plan.
- *4.20.4 Knowledge of Basic Statistical Concepts.* Basic concepts such as variation, control (stability), capability, and overadjustment should be understood throughout the supplier's organization as appropriate.

Customer-specific requirements

Section II of QS-9000 contains individual supplier quality requirements of DaimlerChrysler, Ford, General Motors, and the truck manufacturers that could not be harmonized. Big 3 customer-specific requirements are summarized briefly below.

DaimlerChrysler requirements cover third-party registration; product assurance planning (PAP); the Shield and Diamond symbols to identify special characteristics; significant characteristics; annual layout inspection; frequency of product verification, design validation, and internal quality audits; a seven-step corrective action process; appearance masters; packing, shipping, and labeling; process signoff (PSO); control plans; the Extended Enterprise network; and the supply partner information network (SPIN).

Ford requirements cover third-party registration, control item parts, annual layout, setup verification, control item fasteners, heat treating, process and design changes for supplier-responsible designs, supplier notification of control item requirements, engineering specification (ES) test performance requirements, prototype part quality initiatives, QOS assessment methodology, APQP status reporting guidelines, run at rate, supplier laboratory requirements and calibration services, product qualification, and ongoing process and product monitoring.

General Motors requirements include third-party registration, QS-9000 applicability, UPC labeling for commercial service applications, layout inspection and functional test, control plan signatures, PPAP, year 2000 supplier readiness, electronic communication, shipment notification system, and 17 GM publications containing additional requirements and guidelines.

Customer-specific requirements of Mack Trucks, Inc., Navistar International Transportation Corp., PACCAR, Inc., Volvo Truck North America, Mitsubishi Motors-Australia, and Toyota Australia are set forth in company publications.

TE Supplement

The Tooling and Equipment (TE) Supplement to QS-9000 provides the nuts and bolts needed by 50,000 to 75,000 automotive tooling and equipment suppliers to achieve continuous improvement in their production processes. It was developed by a work group of the Chrysler/Ford/General Motors Supplier Quality Requirements Task Force, with the first edition released in 1996, replacing Chrysler's *Tooling and Equipment Supplier Quality Assurance* (TESQA) manual and Ford's *Facilities and Tools Quality System Standard* (F&T QSS). The substantially revised second edition followed in 1998.

The TE Supplement, which is designed to add value at each stage of the tooling and equipment product life cycle, includes all ISO 9001 and most QS-9000 requirements, along with tooling and equipment quality requirements. It is used with QS-9000 as one document for the tooling and equipment industry. The TE Supplement provides all nonproduction tooling and equipment suppliers with common system guidelines unique to the manufacturing of tooling and equipment, such as tools, dies, molds, plating, robotics, and assembly, along with some coolants and lubricants. As with QS-9000, Element 4.4, Design Control, does not apply to a company that does not have design capability.

The goal of the TE Supplement is to improve the quality, reliability, maintainability, and durability of tooling and equipment supplied to DaimlerChrysler, Ford, and General Motors. This is accomplished by developing and implementing fundamental quality management systems that provide for continuous improvement, emphasizing defect prevention and the reduction of process variation and waste in the tooling and equipment supply chain.

The contents of QS-9000, as interpreted by the TE Supplement, apply to all suppliers of machinery or any component thereof, such as replacement parts. Machinery consists of tooling and equipment to perform processes such as assembling, balancing, casting, deburring, forging, forming, gauging, heat treating, machining, material handling, measuring, molding, packaging, painting, plating, robotics, stamping, tooling, washing, welding, and other emerging manufacturing technologies.

The first edition of the TE Supplement referenced reliability and maintainability requirements from the *Reliability and Maintainability Guideline (for Manufacturing Machinery and Equipment)*. These requirements have been documented in the body of the second edition. Additional requirements may be specified by the customer, and the *Reliability and Maintainability Guideline* may still be considered a good source of additional information.

The *Production Part Approval Process* (PPAP) reference manual does not apply to tooling and equipment suppliers, having been replaced in the TE Supplement by the machinery qualification runoff requirements. The *Advanced Product Quality Planning and Control Plan* (APQP), *Failure Mode and Effects Analysis* (FMEA), *Measurement Systems Analysis* (MSA), and *Fundamental Statistical Process Control* (SPC) reference manuals are available to tooling and equipment suppliers for reference purposes only.

Most QS-9000 customer-specific requirements do not apply to tooling and equipment suppliers. In Section II, both Ford and General Motors include additional company manuals in their TE Supplement customer-specific requirements, whereas DaimlerChrysler sets forth equipment qualification guidelines.

The first edition of the TE Supplement contained no registration provisions. The second edition allows for registration, but only DaimlerChrysler has made it a requirement.

TE supplement sector-specific guidelines

The Big 3 has sprinkled the following sector-specific guidelines throughout the ISO 9000–based requirements in Section I of the TE Supplement.

- *4.1.1 Quality Policy.* Quality objectives must include reliability, maintainability, and durability.
- *4.2.3.1 Advanced Product Quality Planning.* Suppliers must use an advanced quality planning process, embracing reliability and main-

tainability through the life-cycle process. The supplier must fully understand the customer's quality, reliability, maintainability, and durability requirements. The tooling and equipment supplier must document a reliability and maintainability plan that lists techniques, responsibility, and milestone dates to accomplish reliability and maintainability requirements. Implementation of the reliability and maintainability methodologies is intended to be a multidisciplinary functional exercise.

- *4.2.3.2 Special Characteristics.* Controls must be devised and implemented throughout the concept, design and development, prototype, machinery build, and test phases to ensure that quality and reliability requirements are met. These controls must be documented in a control plan.
- *4.2.3.5 Process Failure Mode and Effects Analyses* (*Process FMEAs*). This replaces the QS-9000 Process FMEA section. Use of FMEAs must be documented on all manufactured products. FMEAs, generic to a common process or family of products, must be used when appropriate. The FMEA manual should be used for the proper guidelines to the application and format of FMEAs. Additional guidance may be found in the *Reliability and Maintainability Guideline.* Internal processes should be addressed using this discipline. FMEAs must be living documents and must be reviewed and updated when changes to the original product or process occur, based on performance feedback from the field.
- *4.2.3.7 The Control Plan.* This replaces the QS-9000 Control Plan section. Control plans must be developed by suppliers, using a multidisciplinary approach, to address engineering requirements, special characteristics, and process controls for the manufacture of tooling and equipment. Each piece of machinery must have a control plan, but in many cases, generic control plans can cover tooling and equipment using a common process, such as cutting tools, tool holders, die details, and presses. The control plan format is at the supplier's discretion. The output of the advanced quality planning process, beyond the development of robust processes, is a control plan. Control plans may be based on existing plans for mature product and capable processes. Control plans must be living documents and must be reviewed and updated when changes to the original product or process occur. Customer approval of control plans may be required.
- *4.2.4 Machinery Qualification Runoff Requirements.* This replaces the QS-9000 Product Approval Process section.
- *4.2.4.1 Purpose.* Qualification requirements have been developed to ensure that robots, machinery, tools, and equipment purchased by a

customer will be of acceptable quality when received, in both function and reliability. Specifically, all identifiable problems must be eliminated prior to the machinery being integrated into a larger system or installed at the customer's facility. In addition, this machinery is expected to function properly from delivery throughout its intended service life. Specifically, the intent of this procedure is to reduce or eliminate startup delays, improve the quality of all components and systems to a level that conforms to customer standards, resolve software and control problems prior to launch, confirm that equipment cycle time will meet the customer's productivity requirements, and verify reliability of tooling and equipment.

- *4.2.4.2 Machinery Qualification Runoff Requirements.* Any failure during a machinery qualification runoff must be documented with root cause analysis by the supplier.

- *4.2.4.3.a 50/20 Dry Run.* The 50-hour quality test has been established for robots. At the customer's discretion, the OEM's test data may be used in lieu of an on-site 50-hour dry run. The 20-hour continuous dry run applies to all machinery, including all robots within these systems. All tests must be performed at the system supplier's facility or as directed by the customer. Customer personnel will be on site at the beginning of each test and may provide assistance. The supplier must provide the required resources to run and service the machinery during the test, including maintenance in accordance with the established maintenance schedule. The system supplier must be responsible for all components during all test runs. Time required to comply with this program must be recognized as part of the schedule for delivery. Failure during the 20-hour run requires a restart of the 20-hour requirement, unless waived by the customer. Tooling, such as stamping dies, injection molds, and perishable tooling, may be exempt from the 20-hour continuous dry run requirement.

- *4.2.4.3.b Phase 1 of Initial Process Performance—Preliminary Evaluation.* A sample run will be the first formalized process evaluation with approved parts. This test is primarily for the benefit of the supplier and can be performed without customer representation. However, all parts should be identified by sequence and be available for review by customer personnel.

- *4.2.4.3.c Phase 2 of Initial Process Performance—Pp.* Pp is a measure of process performance irrespective of tolerance location. This evaluation will be conducted at the supplier's location, possibly with a customer representative present. The supplier must consult with the appropriate customer representative for clarification of sample size and type, frequency, total quantity, customer representation requirements, and acceptance criteria. On processes with inherently low variation, target nominals may be purposely shifted from print nominals to favor assembly or allow for tool wear.

- *4.2.4.3.d Phase 3 of Initial Process Performance—Ppk.* Ppk is a measure of process performance and how it relates to tolerance location while considering tolerance width. This evaluation will be conducted at the supplier's location, possibly with a customer representative. If a Ppk calculated from the phase 2 data is found acceptable, then the phase 3 Ppk requirement has been met simultaneously. The supplier must consult with the appropriate customer representative for clarification of sample size and type, frequency, total quantity, customer representation requirements, and acceptance criteria. On processes with inherently low variation of known naturally skewed distribution, target nominals may be purposely shifted from print nominals to favor assembly or allow for tool wear. Decisions on changing dimensional targeting should be agreed to by the entire planning team and only after process capability has been demonstrated.
- *4.2.4.3.e Reliability Verification.* The final part of the predelivery acceptance procedure is a projection of machinery reliability by the supplier. The supplier must demonstrate through documented reliability and maintainability engineering analysis, such as thermal analysis, stress analysis, and reliability predictions, that machinery is designed and built to the specified reliability requirement.
- *4.2.4.3.f 20-Hour Dry Run.* To verify the as-installed condition of the machinery, the 20-hour dry run is to be repeated. As before, the system must cycle at production speeds or a previously agreed to cycle speed continuously for 20 hours. Both customer and supplier(s) should be represented throughout this run.
- *4.2.4.3.g Short-Term Process Study.* This check is to be performed under production conditions. Data from 25 subgroups (recommended 125-piece total) are to be collected and recorded. Specific data summarization may be specified by the customer.
- *4.2.4.3.h Long-Term Process Study.* The 25-working-day (200-hour) long-term process study is conducted at the customer's plant per its requirements. The tooling and equipment supplier should consult its plant representative regarding the requirements of this study.
- *4.2.5.3 Techniques for Continuous Improvement.* The supplier also must demonstrate knowledge of mean time between failures (MTBF), mean time to repair (MTTR), life-cycle cost (LCC), and reliability growth.
- *4.4.2.1 Required Skills.* Additional required skills include mean time to repair (MTTR), mean time between failures (MTBF),

fault-tree analysis (FTA), life-cycle cost (LCC), and environmental characterization.

- *4.4.5.1 Design Output—Supplemental.* Design output also must include analysis of test data, such as thermal, stress, and fault-tree analyses, and projections of reliability, maintainability, durability, and life-cycle cost.
- *4.4.9 Design Changes—Supplemental.* Suppliers must maintain a log of all design changes throughout each phase of the machinery build, including those requested by the customer.
- *4.4.10 Customer Prototype Support.* This replaces the QS-9000 Customer Prototype Support section. Performance tests must consider and include, as appropriate, reliability, maintainability, durability, and product life cycle. When required by the customer, the supplier must have a comprehensive prototype program and/or use predictive reliability and maintainability techniques, such as simulation or mathematical modeling, in order to confirm that tooling and equipment designs meet performance objectives. Accelerated life tests must be conducted on crucial components of machinery or equipment. Crucial components are those having the greatest impact on the reliability of the end product. Test conditions must be accelerated in an appropriate manner so that the failure rate can be translated into normal use conditions.
- *4.6.2.1 Subcontractor Development.* The contents of QS-9000 apply with the exception that subcontractor development must use the TE Supplement as the fundamental quality system requirement.
- *4.8 Product Identification and Traceability.* The supplier must establish and maintain a tracking system for components and subassemblies. An effective system must identify components to their next operation and by their job number. The tracking system also should cross-reference engineering drawings and bills of materials or equivalent.
- *4.9.1 Process Monitoring and Operator Instructions.* This replaces the QS-9000 Process and Monitoring and Operator Instructions section. The use of process monitoring and operator instructions must be adequate to document the tooling and equipment supplier's process. Process monitoring and operator instructions may take the form of process sheets, inspection and laboratory test instructions, shop travelers, test procedures, standard operation sheets, or other documents normally used by the supplier to provide the necessary information. Each employee must understand the work instructions and the objective of his or her job assignment.
- *4.9.2 Maintaining Process Control.* The QS-9000 Maintaining Process Control section applies to high-volume production runs,

such as perishable tooling. *Machinery Qualification Runoff Requirements* replace PPAP.

- *4.10.4.1 Layout Inspection and Functional Testing.* This replaces the QS-9000 Layout Inspection and Functional Testing section. Functional verification to applicable customer engineering material and performance standards must be performed for all tooling and equipment. The *Machinery Qualification Runoff Requirements* summarize the testing required for final buyoff. The results must be available for customer review on request.
- *4.15.6.1 Supplier Delivery Performance Monitoring.* This replaces the QS-9000 Supplier Delivery Performance Monitoring section. The supplier must establish a goal of 100 percent on-time shipments to meet customer requirements. Suppliers must develop timing plans that can be used for management planning and control and which can be adjusted to show the effect of changes when needed. Critical path scheduling must be required for timing control of complex manufacturing systems. Scheduling processes must be in place that accurately control start and end timing for the manufacture of all major components and assemblies for equipment test, runoff, installation, and tryout. These scheduling systems must include subcontractors and correspond with material procurement timing and capacity resource planning. Suppliers should develop resource plans that identify capacity limitations for machines, labor, and facilities and allow alternative plans to be developed when needed. Regular timing meetings should be held to determine the status of actions needed to maintain timing and to determine the effect of engineering changes.
- *4.18 Training.* The supplier must implement a formal training program that includes reliability and maintainability.
- *4.19.1 Feedback of Information from Service.* A procedure for communicating information on machine uptime, availability, reliability, maintenance history, and service concerns to manufacturing engineering and design activities must be established and maintained.
- *4.20.1 Identification of Need.* Statistical techniques also must include reliability and maintainability analysis.
- *4.20.4 Knowledge of Basic Statistical Concepts.* Statistical concepts include mean time between failures (MTBF), mean time to repair (MTTR), short-run statistical process control (SPC), control charts, and p, np, c, and u charts or any other appropriate statistical technique.

The following QS-9000 elements do not apply to tooling and equipment suppliers and are available for reference only.

- *4.9.6 Appearance Items*
- *4.13.3 Control of Reworked Product*
- *4.13.4 Engineering Approved Product Authorization*
- *4.15.6.2 Production Scheduling*
- *4.15.6.3 Electronic Communication*
- *4.15.6.4 Shipment Notification System*

VDA 6.1, EAQF, AVSQ

ISO 9000 serves as the framework for three European automotive quality standards. In Germany, the Verbrand der Automobilindustrie e.V. (VDA), with input from major manufacturers and suppliers, produced VDA 6.1. In France, the Fédération des Industries des Équipements pour Véhicules (FIEV) and the Comité des Constructeurs Français d'Automobiles (CCFA) developed Evaluation Aptitude Qualité Fournisseur (EAQF). In Italy, the Associazione Nazionale Fra Industrie Automobilistiche (ANFIA) created ANFIA Valutazione Sistemi Qualità (AVSQ).

Each of these standards organizes ISO 9000 and automotive industry quality requirements in different ways. VDA 6.1, for example, is organized into 23 elements in two areas: (1) Management and (2) Product and Process. These elements include portions of ISO 9001 and ISO 9004-1, along with automotive sector–specific QS-9000 and EAQF requirements.

VDA 6.1 applies to suppliers of such German automotive manufacturers as Volkswagen, Audi, Mercedes-Benz, BMW, Porsche, Adam Opel, and Ford-Werke. EAQF applies to suppliers of such French automotive manufacturers as Peugeot, Citroën, and Renault. AVSQ applies to suppliers of such Italian automotive manufacturers as Fiat, Ferrari, Lamborghini, and Maserati.

The existence of four automotive quality standards created difficulties for international suppliers, subjecting them to conflicting requirements and multiple registration audits. The first attempt to resolve these problems occurred in 1994, when the German and French automotive industries agreed to mutual recognition of VDA 6.1 and EAQF quality system audit results.

In 1996, a limited reciprocal agreement was reached, under which the automakers behind QS-9000, VDA 6.1, EAQF, and AVSQ recognize each other's internal audit and subcontractor development requirements. This means, for example, that a supplier registered to VDA 6.1 and seeking QS-9000 registration has already fulfilled these two QS-9000 elements.

ISO/TS 16949

The reciprocal agreement, along with the recognition that 85 percent of the four standards consists of common requirements, led to further harmonization discussions and formation of the International Automotive Task Force (IATF). In 1996, IATF and representatives from ISO TC 176 agreed to develop a common and integrated international automotive quality systems standard that would provide for a single third-party registration acceptable to the Big 3 and European automakers. The result of these harmonization efforts is ISO/Technical Specification (TS) 16949, *Quality systems—Automotive suppliers—Particular requirements for the application of ISO 9001: 1994,* released in 1999.

ISO/TS 16949 contains all of ISO 9001, with automotive sector–specific requirements from QS-9000, VDA 6.1, EAQF, and AVSQ added to nearly every element. Big 3 suppliers will find that it is organized along the same lines as QS-9000, with many similar requirements and some differences in emphasis or scope. As with QS-9000, each participating automaker will have customer-specific requirements. Applicable customer-specific requirements will be audited as part of ISO/TS 16949 registration, and the particular automaker(s) will be listed on the supplier's registration certificate.

ISO/TS 16949: 1999 is available as an alternative to the four national automotive standards. However, this arrangement may prove to be temporary, for it will have a short shelf life. When ISO 9001: 2000 is released, ISO/TS 16949 will be revised in the first quarter of 2001 to conform to it. Japanese automakers, who did not participate in drafting ISO/TS 16949: 1999, will take part in this revision. After this process is completed, ISO/TS 16949 may eventually replace all national automotive quality standards.

ISO/TS 16949 sector-specific requirements

The following automotive sector–specific requirements appear among the ISO 9001 elements in ISO/TS 16949.

- *4.1.1.2 Objectives.* Suppliers must develop goals, objectives, and measurements to deploy the quality policy. Objectives for achieving quality must be included in the business plan.
- *4.1.1.3 Customer Satisfaction.* Suppliers must have a documented process to determine customer satisfaction. This must include frequency of determination, how objectivity and validity are ensured, and indicators to monitor customer satisfaction trends.
- *4.1.1.4 Continuous Improvement.* Continuous improvement in quality, service, cost, and technology must be covered in the quality policy. Suppliers must identify opportunities for quality and productivity

improvement, implement appropriate improvement projects, and use appropriate continuous improvement measures and methodologies.

- *4.1.2.1.2 Customer Representative.* Suppliers must assign responsibility to appropriate individuals to represent customer needs in internal functions that address quality requirements.
- *4.1.2.1.3 Quality Responsibility.* Management with corrective action responsibility and authority must be promptly informed of noncompliant products or processes. Quality-responsible personnel have the authority to stop production to correct quality problems.
- *4.1.2.2.2 Shift Resources.* All production shifts must be staffed with personnel responsible for quality.
- *4.1.2.4 Organizational Interfaces.* This is nearly identical to QS-9000 Element 4.1.2.4.
- *4.1.3.2 Management Review—Supplemental.* Management reviews must include all elements of the quality system and its performance over time as an essential part of the continuous improvement process, along with monitoring strategic quality objectives and regular reporting and evaluation of the cost of poor quality
- *4.1.4 Business Plan.* This is nearly identical to QS-9000 Element 4.1.4.
- *4.1.5 Analysis and Use of Company-Level Data.* This is nearly identical to QS-9000 Element 4.1.5, adding a requirement for timely reporting of product usage information.
- *4.1.6 Employee Motivation, Empowerment, and Satisfaction.* Suppliers must have a process to motivate employees to achieve quality objectives and make continuous improvements. This process includes promoting quality awareness on all levels. Suppliers must have a process to measure employee satisfaction and understanding of quality objectives.
- *4.1.7.1 Product Safety.* Due care regarding product safety and minimizing risks to employees, customers, users, and the environment must be addressed in the supplier's quality policy and practices, especially in design control and process control. The supplier must promote internal awareness of product safety considerations.
- *4.1.7.2 Regulations.* The supplier must have a process to ensure compliance with all applicable government, safety, and environmental regulations, including those concerning storage, handling, recycling, eliminating, or disposing of materials.
- *4.2.2.2 Quality System Documentation.* All ISO/TS 16949 requirements must be addressed in the quality system documentation but not necessarily by individual procedures.

- *4.2.3.2 Quality Plan Requirements.* Suppliers must have quality plans that include customer requirements and references to appropriate technical specifications.
- *4.2.4.1 Product Realization—General.* The supplier must have a product realization process to deliver products on time that meet customer quality, cost, delivery, and design requirements. If a project management approach is used, a project manager and a project team will be assigned, appropriate resources will be allocated, and any special responsibilities and organizational interfaces will be defined. The supplier must ensure the confidentiality of customer-contracted products and projects.
- *4.2.4.2 Measurements.* Measurements at appropriate product realization stages must be defined, analyzed, and reported to management. These measurements include quality, risks, costs, lead times, and critical paths.
- *4.2.4.3 Review Cycle.* The supplier must review the status at appropriate product realization stages and take suitable action.
- *4.2.4.4 Multidisciplinary Approach.* The supplier must use a multidisciplinary approach to prepare for product realization, including developing special characteristics, failure mode and effects analyses (FMEAs), and control plans.
- *4.2.4.5 Tools and Techniques.* The supplier must use tools and techniques identified in *Advanced Product Quality Planning and Control Plan* (APQP) customer manuals. These tools and techniques include process FMEAs to analyze potential nonconformities, mistake-proofing methods, and process capability studies on all new processes. Process capability study results must be used to establish production equipment specifications where applicable.
- *4.2.4.6 Computer-Aided Design.* When specified in the contract, the supplier must have the appropriate resources and equipment to use computer-aided product design, engineering, and analysis compatible with customer systems, including subcontracted design work. Where applicable, the supplier must be able to use computer-aided numerical design and drawing data to manufacture precision tooling and prototypes.
- *4.2.4.7 Special Characteristics.* The supplier must apply appropriate methods to identify special characteristics and must include all special characteristics in the control plan.
- *4.2.4.8 Feasibility Reviews.* Suppliers must investigate and confirm the manufacturing feasibility of proposed products in contract review.

- *4.2.4.9.1 Management of Process Design—General.* The supplier must establish and maintain documented procedures to develop and verify process designs used for product realization.
- *4.2.4.9.2 Process Design Input.* The supplier must identify, document, and review the process design input requirements.
- *4.2.4.9.3 Process Design Output.* The process design output must be expressed in terms that can be verified and validated against process design input requirements.
- *4.2.9.4 Process Verification.* The supplier must verify process design output against process design input requirements and record the results.
- *4.2.4.10 Control Plan.* This is nearly identical to QS-9000 Element 4.2.3.7.
- *4.2.4.11 Product Approval Process.* The supplier must comply with a customer-recognized product and process approval procedure. This product approval process also must be applied to subcontractors. The supplier must verify that changes are validated. All changes require customer notification and may require customer approval. When the customer requires, additional verification or identification requirements, such as those for new model introduction, must be met.
- *4.2.5 Plant, Facility, and Equipment Planning.* This is nearly identical to QS-9000 Element 4.2.6.1.
- *4.2.6 Tooling Management.* This is nearly identical to QS-9000 Element 4.2.6.2.
- *4.2.7 Process Improvement.* This is nearly identical to QS-9000 Element 4.2.5.1.
- *4.2.8 Quality System Performance.* The supplier must evaluate quality system performance to verify the effectiveness of its operation. Results must be used for continuous improvement or corrective action as appropriate.
- *4.3.2.2 Contract Review—Supplemental.* Suppliers must have a process to identify cost elements or price, as appropriate, in developing quotations. The supplier must ensure that any customer-specific requirements are met.
- *4.4.2.2 Required Skills.* The supplier must ensure that the design team is qualified to achieve design requirements. The supplier's design team must be qualified in appropriate skills such as geometric dimensioning and tolerancing (GD&T), quality function deployment (QFD), design for manufacturing (DFM)/design for assembly (DFA), value engineering (VE), design of experiments (DOE), failure

mode and effects analysis (DFMEA/PFMEA, etc.), finite-element analysis (FEA), solid modeling, simulation techniques, computer-aided design (CAD)/computer-aided engineering (CAE), and reliability engineering plans.

- *4.4.2.3 Research and Development.* Suppliers must have access to research and development facilities to ensure product and process innovation.
- *4.4.4.2 Reliability Objectives.* Product life, reliability, durability, and maintainability objectives must be included in design inputs.
- *4.4.4.3 Use of Information.* The supplier must have a process to deploy information gained from previous design projects, competitor analysis, or other sources, as appropriate, for current and future projects of a similar nature.
- *4.4.5.2 Design Optimization.* This is nearly identical to QS-9000 Element 4.4.5.1.
- *4.4.8.2 Design Validation—Supplemental.* This is nearly identical to QS-9000 Element 4.4.8.1.
- *4.4.8.3 Prototype Program.* This is nearly identical to QS-9000 Element 4.4.10.
- *4.4.9.2 Evaluation of Design Change.* The supplier must address the impact of a design change on the systems in which the product is used, the customer assembly process, and other related products and systems.
- *4.5.2.2 Engineering Specifications.* This is nearly identical to QS-9000 Element 4.5.2.1.
- *4.6.1.2 Customer-Approved Subcontractors.* Where specified by the contract, suppliers must purchase products, materials, or services from approved subcontractors. Other subcontractors may be used only after they have been approved by the customer. The use of customer-designated subcontractors does not relieve the supplier of the responsibility for ensuring the quality of subcontracted parts, materials, or services.
- *4.6.1.3 Regulatory Compliance.* All purchased products or materials used in parts manufacture must satisfy current regulatory requirements applicable to the country of manufacture and sale, such as environmental, electrical, electromagnetic, and safety.
- *4.6.2.2 Subcontractor Development.* Suppliers must perform subcontractor quality system development, with the goal of subcontractor compliance to ISO/TS 16949 or an existing customer quality system requirements manual.

- *4.6.2.3 Scheduling Subcontractors.* This is nearly identical to QS-9000 Element 4.6.2.1
- *4.7.2 Customer-Owned Tooling.* This is nearly identical to QS-9000 Element 4.7.1.
- *4.9.1.2 Cleanliness of Premises.* This is nearly identical to QS-9000 Element 4.9.b.1.
- *4.9.1.3 Contingency Plans.* This is nearly identical to QS-9000 Element 4.9.b.2.
- *4.9.1.4 Designation of Special Characteristics.* This is nearly identical to QS-9000 Element 4.9.d.1.
- *4.9.1.5 Preventive Maintenance.* This is nearly identical to QS-9000 Element 4.9.g.1.
- *4.9.2 Job Instructions.* The supplier must prepare documented job instructions for all employees having responsibilities for operating processes. These instructions must be accessible for use at work stations without disruption to the job and must be derived from appropriate sources, such as the quality plan, control plan, and product realization process.
- *4.9.3 Maintaining Process Control.* The supplier must maintain or exceed process capability or performance as approved via the customer part approval process. To accomplish this, the supplier must ensure that the control plan and process flow diagram are implemented, including, but not limited to, adherence to the specified measurement technique, sampling plans, acceptance criteria, and reaction plans when acceptance criteria are not met. Significant process events, such as tool change or machine repair, must be noted on the control charts. The supplier must initiate the appropriate reaction plan from the control plan for characteristics that are either unstable or noncapable. These reaction plans must include containment of process output and 100 percent inspection, as appropriate. A supplier corrective action plan must then be completed, indicating specific timing and assigned responsibilities to ensure that the process becomes stable and capable. The plans are to be reviewed with and approved by the customer when so required.
- *4.9.4 Verification of Job Setups.* This is nearly identical to QS-9000 Element 4.9.4.
- *4.9.5 Appearance Items.* This is nearly identical to QS-9000 Element 4.9.6.
- *4.10.1.2 Acceptance Criteria.* Acceptance criteria for attribute data sampling plans must be zero defects. Appropriate acceptance

criteria for all other situations, such as visual standards, must be documented by the supplier.

- *4.10.2.4 Incoming Product Quality.* Unless waived by the customer, the supplier's incoming quality system must use either receipt and evaluation of statistical data by the supplier; receiving inspection and/or testing, such as sampling based on performance; second- or third-party audits of subcontractor sites, when coupled with records of acceptable quality performance; or part evaluation by accredited laboratories.
- *4.10.4.2 Layout Inspection and Functional Testing.* This is nearly identical to QS-9000 Element 4.10.4.1.
- *4.10.6 Laboratory Requirements.* When inspection, testing, and calibration services are conducted by a supplier's laboratory facility, the laboratory must comply with ISO/IEC 17025, including use of a laboratory scope. Commercial or independent laboratories used for inspection, testing, or calibration services by the supplier must be accredited to ISO/IEC 17025 or a national equivalent.
- *4.11.1.2 Measurement System Analysis.* Appropriate statistical studies must be conducted to analyze variations in measuring and test equipment results. This requirement applies to measurement systems referenced in the control plan. The analytical methods and acceptance criteria used must conform to customer measurement system analysis reference manuals as appropriate, including bias, linearity, stability, and measurement repeatability and reproducibility studies. Other analytical methods and acceptance criteria may be used if approved by the customer.
- *4.11.3 Records.* This is nearly identical to QS-9000 Element 4.11.3.
- *4.13.1.2 Suspect Material or Product.* This is nearly identical to QS-9000 Elements 4.13.1.1 and 4.13.1.2.
- *4.13.1.3 Corrective Action Plan.* The supplier must quantify, analyze, and reduce all nonconforming product by establishing a corrective action plan. Progress toward the plan must be tracked. Customers must be informed promptly when nonconforming product has been shipped.
- *4.13.3 Control of Reworked Product.* Rework instructions must be accessible and used by the appropriate personnel in their work areas.
- *4.13.4 Engineering Approved Authorization.* This is nearly identical to QS-9000 Element 4.13.4.

- *4.14.1.2 Problem Solving.* The supplier must use problem-solving methods when an internal or external nonconformity to specifications or requirements occurs. When external nonconformities occur, the supplier must respond in a manner acceptable to the customer. If a customer-prescribed format exists, the supplier must use it.
- *4.14.1.3 Mistake-Proofing.* This is nearly identical to QS-9000 Element 4.14.1.2.
- *4.14.2.2 Corrective Action Impact.* The supplier must apply the corrective action taken and the controls implemented to eliminate the cause of a nonconformity to other similar processes and products.
- *4.14.2.3 Returned Product Test/Analysis.* The supplier must analyze parts returned from customers' manufacturing plants, engineering facilities, and dealerships. The supplier must minimize the cycle time of this process. Records of these analyses must be kept and made available on request. The supplier must perform effective analysis and, where appropriate, initiate corrective action to prevent recurrence.
- *4.15.3.2 Inventory.* The supplier must use an inventory management system to optimize inventory turns over time and ensure stock rotation, such as first in, first out (FIFO). Obsolete product must be controlled in a similar manner to nonconforming product.
- *4.15.4.2 Customer Packaging Standards.* The supplier must comply with all customer packaging requirements, including those applicable to service parts.
- *4.15.4.3 Labeling.* This is nearly identical to QS-9000 Element 4.15.4.2
- *4.15.6.2 Performance Monitoring of Supplier Delivery.* The supplier must establish systems to support 100 percent on-time deliveries to meet customer production and service requirements. If 100 percent on-time deliveries are not maintained, the supplier must inform the customer of an anticipated delivery problem and implement corrective actions to improve delivery performance. The supplier must have a systematic approach to develop, evaluate, and monitor adherence to established lead-time requirements. The supplier must implement a system to monitor performance to customer delivery requirements, with corrective actions taken as appropriate. Records of supplier-responsible premium freight must be maintained. The supplier must ship all product or materials in conformance with customer requirements, adhering to up-to-date customer-specified transportation mode, routings, and containers.
- *4.15.6.3 Production Scheduling.* There must be appropriate production scheduling to meet customer requirements, such as just-in-

time, supported by an information system that permits access to production information at key process stages and is order-driven.

- *4.15.6.4 Electronic Communication.* This is nearly identical to QS-9000 Element 4.15.6.3.
- *4.15.6.5 Shipment Notification System.* This is nearly identical to QS-9000 Element 4.15.6.4.
- *4.16.2 Record Retention.* The supplier must define retention periods for quality system–related documents and records to satisfy regulatory and customer requirements as a minimum.
- *4.17.2.1 Internal Quality Audits—Supplemental—General.* When internal or external nonconformities or customer complaints occur, the audit frequency must be increased appropriately.
- *4.17.2.2 System Audit.* Internal system audits must cover all activities and shifts. These audits must be scheduled according to an annual plan to verify compliance with ISO/TS 16949 and any additional system requirements.
- *4.17.2.3 Process Audit.* The supplier must audit the product realization and production processes to determine the effectiveness of process performance.
- *4.17.2.4 Product Audit.* The supplier must audit products at appropriate stages of production and delivery to verify conformance to specific requirements, such as product dimensions, packaging, and labeling, at an appropriate frequency.
- *4.17.2.3 Auditor Qualification.* The supplier must comply with customer requirements for internal system and product auditor qualification.
- *4.18.2 Training Effectiveness.* Training effectiveness must be reviewed periodically. Special attention must be given to customer-specific requirements.
- *4.18.3 Training on the Job.* The supplier must provide on-the-job training for personnel, including contract or agency personnel, in any new or modified job affecting quality, if appropriate. Personnel in jobs affecting quality must be informed about the consequences for the customer of quality standard nonconformities.
- *4.19.2 Feedback of Information from Service.* This is nearly identical to QS-9000 Element 4.19.1.
- *4.19.3 Servicing Agreement with Customer.* When there is a servicing agreement with the customer, the supplier must verify the effec-

tiveness of any supplier service centers, special-purpose tools, equipment, and training of servicing personnel.

- *4.20.3 Identification of Statistical Tools.* Appropriate statistical tools for each process must be determined during advance quality planning and included in the control plan.
- *4.20.4 Knowledge of Basic Statistical Concepts.* This is nearly identical to QS-9000 Element 4.20.4.

Chapter

17

Other ISO 9000 Derivatives

ISO 9000 is a generic standard that is adaptable to any industry. As the preceding chapter demonstrated, however, many industries have quality needs that go beyond ISO 9000 requirements. The solution to this problem is developing sector-specific ISO 9000 derivatives. As with the automotive standards examined in the preceding chapter, these derivatives combine ISO 9001 with additional requirements for the particular industry.

It should be no surprise therefore that other business sectors have followed the automotive industry's lead by developing their own ISO 9000 derivatives. As with the automotive industry, this process began with individual company quality standards for suppliers, followed by development of national and then international sector-specific ISO 9000 derivatives.

Following processes similar to the automotive industry, the aerospace industry has developed AS9100, while the telecommunications field has produced TL 9000, two of the most prominent ISO 9000 derivatives. Where there is overwhelming industry support, other sector-specific derivatives are likely to follow. This chapter examines AS9100 and TL 9000.

AS9100

In 1995, the American Aerospace Quality Group (AAQG), a committee of procurement, purchasing, and quality executives from the prime U.S. aerospace contractors, began work on an industry-specific quality management system that could save their corporations and 20,000 to 30,000 suppliers time and money. MIL-Q-9858 and MIL-I-45208, for decades the two most common standards in the aerospace industry, had been canceled, leaving no comprehensive quality standard to fill the vacuum. Since military specifications had been slowly giving way to quality standards based on ISO 9000, the time was right to develop an aerospace derivative.

The AAQG operated as a subcommittee under the Aviation/Space and Defense Division of the American Society for Quality (ASQ) and worked in conjunction with the Aerospace Industries Association (AIA) and the General Aviation Manufacturers Association (GAMA), a small-jet trade group that includes Cessna and Learjet. The major companies involved in this effort were McDonnell-Douglas Corporation, The Boeing Company, Lockheed-Martin, Northrop-Grumman, Allison Engine Company, AlliedSignal Aerospace, Pratt & Whitney Aircraft, General Electric Aircraft Engines (GEAE), Rockwell-Collins, Sikorsky Aircraft, and Sundstrand.

In 1996, after less than 18 months of work, a draft standard, SAE Aerospace Resource Document ARD9000, was published by the Society of Automotive Engineers (SAE), the world's largest generator of aerospace industry standards. ARD9000 contained 45 to 50 industry-specific requirements and gained favorable reaction among prime contractors, their suppliers, and the federal government.

Prime contractor executives reviewed the document in order to fine-tune and reduce the number of sector-specific requirements. ARD9000 subsequently was revised, renamed SAE Aerospace Basic Quality System Standard AS9000, and then released as a new standard in 1997.

In developing AS9000, the AAQG took ISO 9001, the Federal Aviation Administration (FAA) Aircraft Certification System Evaluation Program, and Boeing's massive D1-9000 variant of ISO 9000. Members had to balance the need for a standard tailored to aerospace while still generic enough for suppliers to use flexibility and innovation. According to a committee member, too many supplier requirements could seriously impede continuous improvement.

AS9000 contained ISO 9001 in its entirety, along with 27 clarifications or qualifiers and eight notes to the 20 ISO 9001 elements. It was designed to reduce defects in the supplier chain, continually improve quality, boost customer satisfaction, and considerably reduce the number of hours spent on audits, requirements, and documentation.

AS9000 also addressed the flowdown of quality system requirements from aerospace prime contractors to their suppliers and on to supplier subcontractors. This is similar to the subcontractor quality system development requirements in QS-9000 and ISO/TS 16949. Flowdown includes specifications for parts or assembly designs, characteristics, inspections, and other process functions and product features.

AS9000 served as a single American aerospace standard, with flowdown supported and stressed. A supplier previously could face a huge compliance burden, with FAA, Department of Defense (DoD), and MIL-Q standards, as well as private-sector requirements from each customer. For example, an AlliedSignal plant has an average of 23 customers, and at one point, each had its own quality system requirements.

AS9000 did not provide for registration, only qualification. An aerospace supplier could become qualified under AS9000 and register to ISO 9000 and was required to do so by such prime contractors as GEAE and AlliedSignal.

While AS9000 was actively embraced by the aerospace field, as well as the U.S. government, efforts to develop an international aerospace quality standard got under way shortly after its release. An ISO 9000–based European aerospace quality standard, prEN 9000-1, had been developed by the European Association of Aerospace Industries (AECMA). It contained a different set of sector-specific requirements, including configuration management and risk management. As with the automotive industry, international suppliers had to conform to multiple standards.

Working Group 11 of ISO Technical Committee 20, Aircraft and Space Vehicles, consisting of representatives from the AAQG, AECMA, the Society of Japanese Aerospace Companies (SJAC), and the Brazilian aerospace industry, began developing an international aerospace quality standard in 1997. In addition to the companies involved in developing AS9000, such major European aerospace prime contractors as Airbus, British Aerospace, Rolls Royce Allison, and Aerospatiale took part in this effort. In 1998, participating aerospace industry companies, along with SAE, AECMA, and SJAC, formed the International Aerospace Quality Group (IAQG) to achieve significant quality improvements and cost reductions and take responsibility for the new standard's technical contents.

In 1999, SAE Aerospace Standard AS9100, *Quality Systems—Aerospace—Model for Quality Assurance in Design, Development, Production, Installation and Servicing,* an international aerospace quality standard, was released, along with its identical European counterpart, AECMA prEN 9100. AS9100 contains all of AS9000, along with most of prEN 9000-1 and additional requirements submitted by Boeing, the world's largest aerospace manufacturer.

Prior to its release, Boeing, Rolls Royce Allison, Pratt & Whitney, and GEAE announced that they will use AS9100 as the basic quality management system for their suppliers. Boeing also announced that it will use AS9100 for its internal quality management system. A registration system is yet to be developed.

The following aerospace sector-specific requirements appear in AS9100:

- *4.1.2.3 Management Representative.* The management representative must have the necessary authority and organizational freedom to resolve matters pertaining to quality.
- *4.1.2.4 Process Performer.* Suppliers having a quality assurance activity performed by an individual process performer, such as an operator, buyer, or planner, must have procedures that define the specific tasks and responsibilities that are authorized and the corresponding requirements and training necessary to perform those tasks.
- *4.2.1 Quality System—General.* Other quality system requirements imposed by the applicable regulatory authorities must be included or referenced in the quality system documentation.
- *4.2.2.c Quality System Procedures.* The supplier must ensure that quality system procedures are readily accessible to personnel who are responsible for performing work in conformance with requirements and to customer and/or regulatory authorities representatives.
- *4.2.3 Quality Planning.* The supplier must give consideration to
  - *b.* The design, manufacture, and use of tooling so that variable measurements can be taken, particularly for key characteristics.
  - *f.* The identification of in-process verification points when adequate verification of conformance cannot be performed at a later stage of realization.
  - *i.* The identification and selection of subcontractors.
  - *j.* The establishment of appropriate process controls and development of control plans where key characteristics have been identified.
  - *k.* The identification of material, processes, and services to support operation and maintenance of product.
- *4.2.4 Configuration Management.* The supplier must establish, document, and maintain a configuration management process appropriate to the product.
- *4.3.1 Contract Review—General.* The supplier also must establish and maintain documented procedures for tender review and for the coordination of these activities.
- *4.3.2.d Review.* Before submission of a tender or acceptance of a contract or order, it must be reviewed by the supplier to ensure that risks associated with new technology and/or short delivery time scale have been evaluated.
- *4.3.3 Amendment to a Contract.* Contract review requirements also must apply to contract amendments.

- *4.4.1 Design Control—General.* The responsibilities and authorities for design data approval must be defined. When the supplier subcontracts design or development activities, the supplier must control the subcontracted activity consistent with Element 4.4 requirements.
- *4.4.2.1 Design and Development Management Planning.* The supplier must plan the different phases used to carry out the design and development with respect of the organization, task sequence, mandatory steps, significant stages, and method of configuration control. As appropriate, the supplier must give consideration to structuring the design effort into significant elements according to complexity and, for each element, analyzing the tasks and necessary resources for its design and development. This analysis must consider an identified responsible person, design content, planning constraints, and performance conditions.
- *4.4.2.2 Reliability, Maintainability, Safety.* The different design and development tasks to be carried out must be defined according to specified safety or functional objectives of the product in accordance with customer and/or regulatory authority requirements.
- *4.4.4 Design Input.* The input data to the design must be defined and documented in terms of functional requirements in the case of a product requiring design and development planning. The supplier must establish the input data specific to each element and must review to ensure consistency with requirements.
- *4.4.5 Design Output.* All pertinent data required to allow the product to be identified, manufactured, inspected, used, and maintained must be defined by the supplier, including drawings, part lists, and specifications; a listing of those drawings, part lists, and specifications necessary to define the product's configuration and design features; and information on material, processes, type of manufacturing, and assembly of the product necessary to ensure the product's conformity.
- *4.4.6 Design Review.* Consideration must be given to the validity of design in relation to the design stage objectives, actions that need to be taken in the event of any identified deviation, and decisions necessary for progression to the next stage
- *4.4.8.1 Documentation of Design Verification and Validation.* At the completion of development, the supplier must ensure that reports, calculations, test results, and other data demonstrate that the product definition meets the specification requirements for all identified operational conditions and that the product will function correctly.
- *4.4.8.2 Design Verification and Validation Testing.* Where tests are necessary for verification and validation, these tests must be planned,

controlled, reviewed, and documented to ensure and prove that test plans or specifications identify the product being tested and the resources being used, define test objectives and conditions, parameters to be recorded, and relevant acceptance criteria; test procedures describe the method of operation, the performance of the test, and the recording of the results; the correct configuration standard of the product is submitted for the test; the requirements of the test plan and the test procedures are observed; and the acceptance criteria are met.

- *4.4.9 Design Change Approval.* The supplier's design control system must provide for customer and/or regulatory approval of changes, when required by contract or regulatory requirement.
- *4.5.3 Document Change Incorporation.* The supplier must establish a process to ensure the timely review, distribution, implementation, and maintenance of all authorized and released drawings, standards, specifications, planning, and changes. The supplier must maintain a record of change incorporation and, when required, must coordinate these incorporations with the customer and/or regulatory authority.
- *4.6.1 Purchasing—General.* The supplier must be responsible for the quality of all products purchased from subcontractors, including customer-designated sources.
- *4.6.2 Evaluation of Subcontractors.* The supplier must
  - *d.* Ensure where required that both it and all subcontractors use customer-approved special process sources.
  - *e.* Ensure that the organization having responsibility for approving subcontractor quality systems has the authority to disapprove the use of sources.
  - *f.* Periodically review subcontractor performance. Records of these reviews must be maintained and used as a basis for establishing the level of supplier controls to be implemented.
  - *g.* Maintain procedures that define the necessary actions to take when dealing with subcontractors that do not meet requirements. A list of approved subcontractors must be maintained and must specify the scope of approval.
- *4.6.3 Purchasing Data.* Purchasing documents must contain data clearly describing the product ordered, including where applicable
  - *d.* Design, test, examination, inspection, and customer acceptance requirements and any related instructions and requirements.
  - *e.* Right of access by the purchaser, the customer, and regulatory authorities to all facilities involved in the order and all applicable quality records.

- *f.* Requirements for test specimens, such as production method, number, and storage conditions, for design approval, inspection, investigation, or auditing.
- *g.* Requirements relative to the notification of anomalies, changes in definition, and the approval of their processing.
- *h.* Requirements to flowdown to subtier suppliers the applicable requirements in the purchasing documents, including key characteristics where required.

- *4.6.4 Verification of Purchased Product.* The supplier must implement procedures to verify purchased products. These may include obtaining objective evidence of product quality from subcontractors, through accompanying documentation, conformity certificates, test reports, statistical records, process control, or other means; inspection and audit at source; review of the required documentation; inspection of products at delivery; and delegation of verification to the subcontractor, or subcontractor certification. When delegation is used, the supplier must define delegation requirements and maintain a list of delegations.

- *4.8 Product Identification and Traceability.* According to the level of traceability required by contract or regulatory or other established requirement, the supplier's system must provide for identification to be maintained throughout the product life; all the products manufactured from the same batch of raw material from the same manufacturing batch to be traced, as well as the destination (delivery, scrap) of all products of the same batch; for an assembly, the identity of its components and those of the next higher assembly to be traced; and for a given product, a sequential record of its production (manufacture, assembly, inspection) to be retrieved. The supplier must maintain the identification of the configuration of the product in order to identify any differences between the actual and agreed configurations.

- *4.9.1 Process Control—General.* Controlled conditions must include
  - *b.* A suitable working environment in regard to such matters as temperature, humidity, lighting, and cleanliness.
  - *d.* Monitoring and control of key characteristics where required by purchase order or contract.
  - *h.* Accountability for all product during manufacture, such as part quantities, split orders, and nonconformities.
  - *i.* Evidence that all manufacturing and inspection operations have been completed as planned or as otherwise documented and authorized.
  - *j.* Provision for the prevention, detection, and removal of foreign objects.

- *k.* Utilities and supplies such as water, compressed air, electricity, and chemical products to the extent they affect product quality.

- *4.9.1.1 Product Documentation.* Production operations must be carried out in accordance with approved data. These data must contain as necessary drawings, parts lists, process flowcharts including inspection operations, production documents (manufacturing plans, travelers, routers, work orders, process cards), and inspection documents; a list of specific or nonspecific tools and numerical control (NC) machine programs; and documents associated with specific tools that enable the tools to be designed, produced, validated, controlled, used, and maintained.

- *4.9.1.2 Control of Production Process Changes.* Persons required to approve production process changes must be identified and authorized. The supplier must identify those changes which require customer acceptance under contractual requirements before putting them into effect. Changes affecting processes, production equipment, tools, and programs must be documented. Procedures must be available to control their implementation. The results of changes to production processes must be assessed to confirm that the desired effect has been achieved without adverse effects on product quality.

- *4.9.1.3 Control of Production Equipment, Tools, and Numerical Control (NC) Machine Programs.* Production equipment, tools, and programs must be validated prior to use, maintained, and inspected periodically according to documented procedures. Validation prior to production use must include verification of the first article produced to the design data or specification. Storage requirements, including periodic preservation and condition checks, must be established for production equipment or tooling in storage.

- *4.9.1.4 Control of Work Occasionally Performed Outside the Supplier's Facilities.* When planning to carry out work at a location other than its normal facilities, the supplier must define the procedure to validate the location and control the work.

- *4.9.2 Special Processes.* When production operations call for special processes, these special processes must be identified and qualified before being implemented; the supplier must control applicable aspects, as defined by the process specifications, including special process changes; and the supplier must define the significant operations and parameters in the process to be controlled during production.

- *4.10.1 Inspection and Testing—General.* These procedures must specify the resources and methods to be implemented and methods of recording the results. These procedures must include identification of authorized personnel, limits of authorization, and training and qualification requirements. Inspection documentation must be maintained and controlled by the supplier. This may be part of the manufacturing documentation but must include criteria for acceptance and rejection, the sequence for performing inspection and testing operations, records of inspection results, identification of production inspection instruments, and documents enabling specific inspection instruments to be designed, produced, validated, controlled, used, and maintained. When the supplier subcontracts inspection or test activities, it must control the subcontracted activity consistent with the requirements of Element 4.6.

- *4.10.2.4.* When certification test reports are used to accept material, the supplier must assure that the data in said reports are acceptable per applicable specifications. The supplier must periodically validate test reports.
- *4.10.5 Inspection and Test Records.* Test records must show test results data when required by specification or acceptance test plan. When required to demonstrate product qualification, the supplier must ensure that quality records provide evidence that the product meets the defined requirements.
- *4.10.6 First Article Inspection.* The supplier's system must provide a process, as appropriate, for the inspection, verification, and documentation of the first production article. First article inspection documentation must be retained and must include a list of the characteristics required by the design data and any required tolerances, the results, and when testing is required, test results. The first article inspection must be updated to include production process or configuration changes.
- *4.11.1.1 Control of Inspection, Measuring, and Test Equipment—General.* Responsibilities must be defined regarding the control of inspection, measuring, and test equipment, including those used by operators as well as, where appropriate, test devices and tools supplied by the customer.
- *4.11.2 Control Procedure.* The supplier must
  - *b.* Maintain a list of inspection, measuring, and test equipment, including, where appropriate, test devices and tools supplied by the customer.
  - *f.* When the assessment indicates that the product may be nonconforming, dispose of the nonconformance.
  - *j.* Define the method for recall of measuring devices that require calibration.

- *4.12.1 Authorized Personnel.* Records must identify personnel authorized to verify, certify, and release products.
- *4.12.2 Acceptance Authority Media.* When acceptance authority media are used, such as stamps, electronic signatures, or passwords, the supplier must establish and document controls for the media.
- *4.13.1 Control of Nonconforming Product—General.* The supplier's procedures must take into account process nonconformity that may result in product nonconformity. The term *nonconforming product* includes those products returned by a customer.
- *4.13.2 Review and Disposition of Nonconforming Product.* The supplier's documented procedures must define the process for approving personnel making material review decisions.
- *4.13.2.1 Material Review Authority.* Notwithstanding the requirements of Element 4.13.2, the supplier must not use dispositions of use-as-is or repair, unless specifically authorized by the customer if (1) the product is produced to customer design or (2) the nonconformity results in a departure from the contract requirements. Unless otherwise restricted in the contract, supplier-designed product that is controlled by a customer specification may be dispositioned by the supplier as use-as-is or repair, provided the nonconformity does not result in a departure from customer-specified requirements.
- *4.13.2.2 Regrading Material.* Product dispositioned for regrade requires a change in product identification to preclude the product's original use. Adequate test reports and certifications must reflect the regrading.
- *4.13.2.3 Scrap Material.* Product dispositioned for scrap must be conspicuously and permanently marked, or positively controlled, until physically rendered unusable.
- *4.13.2.4 Notification.* The supplier's system must provide for timely reporting of nonconformities that may affect product already delivered, including any continuing airworthiness actions. Notification must include a clear description of the nonconformance, which includes as necessary parts affected, customer and/or supplier part numbers, quantity, and date(s) delivered.
- *4.14.2 Corrective Action.* The procedures for corrective action must include
  - *e.* Flowdown of the corrective action requirement to a subcontractor, when it is determined that the subcontractor is responsible for the root cause.
  - *f.* Specific actions where timely and/or effective corrective actions are not achieved.

- *4.15.1 Handling, Storage, Packaging, Preservation, and Delivery—General.* These procedures must also cover the specific requirements for cleaning; prevention, detection, and removal of foreign objects; special handling for sensitive products; marking and labeling, including safety warnings; shelf-life control and stock rotation; and hazardous materials where applicable in accordance with product specifications and/or applicable regulations.
- *4.15.6 Delivery.* The supplier must ensure that the accompanying documents for the product are present at delivery as specified in the contract or order and are protected against loss and deterioration.
- *4.16 Control of Quality Records.* Records must be available for review by regulatory authorities as required.
- *4.17 Internal Quality Audits.* The supplier must conduct internal quality audits that assess compliance to its quality system and AS9100 requirements. A flowdown of the requirements from AS9100 through the supplier's quality manual to the working-level procedures must be shown. Detailed tools and techniques must be developed, such as checksheets, process flowcharts, or any similar method to support procedural requirements audits. The acceptability of the selected tools will be measured against the effectiveness of the internal audit process and overall supplier performance. Supplier personnel carrying out these audits must have received appropriate training.
- *4.18 Training.* Training to achieve and maintain awareness and understanding of relevant procedures and instructions must be provided.
- *4.19 Servicing.* Servicing procedures must contain a method of collecting and analyzing in-service data; actions to be taken when problems are identified after delivery, including investigation and reporting activities, and actions on service information consistent with contractual and/or regulatory requirements; the control and updating of technical information; the approval, control, and use of repair schemes; and the controls required for off-site work, such as supplier work undertaken at customer facilities.
- *4.20.2 Statistical Techniques—Procedures.* According to the nature of the product, and depending on the criticality and the specified requirements, these statistical techniques may be used to support design verification, such as reliability, maintainability, and safety; process control, including selection and inspection of key characteristics, process capability measurements, statistical process control (SPC), and design of experiments (DOE); inspection, such as matching the sampling rate to the criticality of the product and to the

process capability; quality management, such as determining required improvement activities; and failure mode and effects analysis (FMEA). When the supplier uses sampling inspection as a means of product acceptance, the plan must be statistically valid and appropriate for use. The plan must preclude the acceptance of known defectives in the lot. When required, the plan must be submitted for customer approval.

TL 9000

Seeking continuous improvement in the quality and reliability of telecommunications service, a group of leading telecommunications service providers (TSPs), including Bell Atlantic, BellSouth, Pacific Bell, and Southwestern Bell, formed the Quality Excellence for Suppliers of Telecommunications (QuEST) Leadership Forum in 1996. The QuEST Forum's goal was to develop a consistent set of worldwide telecommunications industry quality system requirements for hardware, software, and services, along with leading-edge performance-measuring tools.

Establishment of the QuEST Forum was the first time the telecommunications industry banded together to develop quality system requirements. Its membership now consists of more than 60 telecommunications service providers and suppliers, including all Regional Bell Operating Companies (RBOCs), AT&T, GTE, Bell Canada, and such leading telecommunications suppliers as Alcatel, Fujitsu Network Communications, Lucent Technologies, Motorola, Nortel Networks, PairGain Technologies, and Siemens Telcom Network.

Globalization of the telecommunications industry, increased customer and purchaser expectations, and more intense competition had created a need for both service providers and suppliers to implement common quality system requirements and metrics. Better products had to be brought to market faster, with fewer problems and full supplier support.

At the same time, the cost of poor quality (COPQ) had been estimated at $750 million a year in the United States and $10 to $15 billion a year worldwide. Some 25 Bellcore quality and reliability standards had been inconsistently used in the industry for years. ISO 9001 failed to meet telecommunications needs for reliability and associated costs, software development, services, and continuity. Suppliers seeking to conform to both the ISO 9000 and Bellcore standards faced a confusing maze of overlapping requirements and audits that escalated the cost of doing business.

In 1997, the QuEST Forum began work on a new standard to harmonize telecommunications quality system requirements for the design, development, production, delivery, installation, and maintenance of hardware, software, and services. This new standard, which became TL 9000, was intended to consolidate ISO 9000, Bellcore, and other quality requirements; make conformance easier to achieve and demonstrate; reduce redundancy and paperwork; decrease time to market; and improve the total cost of ownership throughout the supply chain.

TL 9000 was released in 1999. Its goals are to foster quality systems that effectively and efficiently protect the integrity and use of telecommunications products, including hardware, software, and services; establish and maintain a common set of quality system requirements; reduce the number of telecommunications quality system standards; define effective cost- and performance-based metrics to guide progress and evaluate results of quality system implementation; drive continuous improvement; enhance customer-supplier relationships; and leverage industry conformity assessment processes.

TL 9000's structure consists of five levels of quality system requirements and metrics. They are ISO 9001; common telecommunications industry quality system requirements (QSRs); hardware, software, and services specific quality system requirements; common telecommunications industry metrics; and hardware, software, and services specific quality system metrics.

There are 83 sector-specific requirements, divided into six categories and marked accordingly: common (C), hardware (H), software (S), services (V), hardware and software (HS), and hardware and services (HV). There are no services and software (VS) elements. A supplier is only required to implement the additional elements that fit its scope of operations. Only a hardware and software company that offers a service must implement all 83 requirements.

Sources of the sector-specific requirements are the QuEST Forum; three Bellcore standards: GR-1202-CORE, *Generic Requirements for a Customer Sensitive Quality Infrastructure,* GR-1252-CORE, *Quality System Generic Requirements for Hardware,* and TR-NWT-000179, *Quality System Generic Requirements for Software*; ISO/IEC 12207, *Information Technology Software Life Cycle Processes*; ISO 9000-3; and ISO 9004-2.

These sector-specific requirements include an additional Element 4.21, Quality Improvement and Customer Satisfaction, which covers such matters as the quality improvement program (4.21.1), customer-supplier relationship (4.21.2), and customer satisfaction (4.21.3.C.1). Other significant provisions address reliability, cost, software development, life-cycle management, and specialized service functions.

TL 9000 breaks ground by establishing cost- and performance-based metrics that measure the reliability and quality performance of hardware, software, and services. *Metrics* are performance-measurement tools that are important from the customer's point of view and include hardware return rates, system outages, number of problem reports, software update quality, on-time delivery, invoice accuracy, and the efficiency and level of success of the supplier's business processes and activities.

Metrics can quantify the benefits gained, assess progress in quality maturity, identify areas where the quality process improvement will have the greatest cost effect, and provide comparative benchmarking capabilities for the industry. Their supplier performance measurements can drive improvements that directly affect the customer and help reduce the annual COPQ within the telecommunications industry.

Metrics link quality improvement to business issues and help senior corporate executives appreciate the value of quality management systems. TSPs can use metrics to determine the financial effects of a supplier's performance on their own operations by applying cost factors to performance measurements.

Metrics are to be reported to the American Society for Quality (ASQ), the QuEST Forum administrator, and will be used for benchmarking. They are also reportable to clients if required by contract. Unlike traditional cost-of-quality (COQ) measures, metrics focus on what immediately affects the customer.

TL 9000 applies to an estimated 10,000 telecommunications suppliers. The United States has a $250 billion a year telecommunications market, with worldwide purchases from the industry's suppliers estimated at more than $125 billion by 200 TSPs.

Expected benefits of TL 9000 are continuous improvement of service to subscribers, enhanced customer-supplier relationships, standardization of quality system requirements, efficient management of external audits and site visits, uniform cost- and performance-based metrics, overall cost reduction and increased competitiveness, enhanced management and improvement of supplier performance, and industry benchmarks for TL 9000 metrics.

Much of the telecommunications industry is already registered to ISO 9000. An upgrade to TL 9000 registration would incorporate the additional requirements into existing quality programs. A company, organizational unit, facility, or product line can be TL 9000 registered in the areas of hardware, software, services, or any combination thereof. Registration may be required contractually by RBOCs.

The QuEST Forum is actively seeking support from the international telecommunications community for worldwide acceptance of TL 9000. Along these lines, it has agreed to form a joint working committee with EIRUS, a European telecommunications consortium created to introduce and manage two European Bellcore standards, *European In-Process Quality Metrics* (E-IPQM) and *European*

Reliability and Quality Measurements for Telecommunications Systems (E-RQMS).

The following telecommunications sector–specific requirements appear in TL 9000. As mentioned earlier, elements are marked common (C), hardware (H), software (S), services (V), hardware and software (HS), and hardware and services (HV).

- *4.1.1.C.1 Quality Objectives.* Objectives for quality include targets for the TL 9000 metrics defined in the TL 9000 *Quality System Metrics* handbook.
- *4.2.2.C.1 Life-Cycle Model.* The supplier must establish and maintain an integrated set of guidelines that covers the life cycle of its products. This framework must contain the processes, activities, and tasks involved in developing, operating, maintaining, and if required, disposing of products spanning the product life.
- *4.2.2.S.1 Support Software and Tools Management.* Suppliers must ensure that internally developed support software and tools used in the product life cycle, such as design and development tools, testing tools, configuration management tools, and documentation tools, are subject to the appropriate quality method(s).
- *4.2.3.C.1 Customer Involvement.* The supplier must establish and maintain methods for soliciting and considering customer input for quality planning activities. Consideration should be given to establishing joint customer-supplier quality improvement programs.
- *4.2.3.C.2 Long- and Short-Term Planning.* The supplier's quality planning activities must include long- and short-term plans with goals for improving quality and customer satisfaction.
- *4.2.3.C.3 Subcontractor Input.* The supplier must establish and maintain methods for soliciting and using subcontractor input for quality planning activities.
- *4.2.3.C.4 Disaster Recovery.* The supplier must establish and maintain methods for disaster recovery to ensure the ability to recreate and service the product throughout its life cycle.
- *4.4.1.C.1 Requirements Traceability.* Suppliers must establish and maintain a method to trace specified requirements through design and test.
- *4.4.2.C.1 Project Plan.* The supplier must establish and maintain a project plan based on the defined product life-cycle model. The plan must include project organizational structure; project roles and responsibilities; interfaces with internal and external organizations; means for scheduling, tracking, issue resolution, and reporting; budgets,

staffing, and schedules associated with project activities; method(s), standards, documented procedure(s), and tools to be used; references to related plans, such as development, testing, configuration management, and quality; project-specific environment and physical resource considerations, such as development, user documentation, testing, and operation; customer, user, and subcontractor involvement during the product life cycle, such as joint reviews, informal meetings, and approvals; management of project quality; risk management and contingency plans, such as technical, cost, and schedules; performance, safety, security, and other critical requirements; project-specific training requirements; required certifications; proprietary, usage, ownership, warranty, and licensing rights; and postproject analysis.

- *4.4.2.C.2 Test Planning.* Test plans must be documented and the results recorded.
- *4.4.2.C.3 End-of-Life Planning.* Suppliers must establish and maintain documented procedures for the discontinuance of manufacturing and/or support of a product by the operation and service organizations.
- *4.4.2.S.1 Estimation.* Suppliers must establish and maintain a method for estimating and tracking project factors during project planning, execution, and change management.
- *4.4.2.S.2 Computer Resources.* Suppliers must establish and maintain methods for estimating and tracking critical computer resources for the target computer.
- *4.4.2.S.3 Integration Planning.* Suppliers must develop and document a plan to integrate software components into the product.
- *4.4.2.S.4 Migration Planning.* When a system or software product is planned to be migrated from an old to a new environment, the supplier must develop and document a migration plan.
- *4.4.4.C.1 Customer and Subcontractor Input.* Suppliers must establish and maintain methods for soliciting and using customer and subcontractor input during the development of new or revised product requirements.
- *4.4.4.C.2 Design Requirements.* Design requirements must be defined and documented and should include quality and reliability requirements; product functions and capabilities; business, organizational, and user requirements; safety, environmental, and security requirements; installability, usability, and maintainability requirements; design constraints; and testing requirements.
- *4.4.4.H.1 Content of Requirements.* The design requirements must include, but are not limited to, nominal values and tolerances, maintainability needs, and end-item packaging requirements.

- *4.4.4.S.1 Identification of Software Requirements.* The supplier must determine, analyze, and document the software requirements of the system.
- *4.4.4.S.2 Requirements Allocation.* The supplier must document the allocation of the product requirements to the product architecture.
- *4.4.5.S.1 Design Output.* The required output from the design activity must be defined and documented in accordance with the chosen method.
- *4.4.5.V.1 Services Design Output.* The required output from the services design must contain a complete and precise statement of the service to be provided.
- *4.4.8.H.1 Periodic Retesting.* The supplier must establish and maintain documented procedures that ensure that products are retested periodically to assess their ability to continue to meet design requirements.
- *4.4.8.H.2 Content of Retesting.* The initial test and periodic retest must be more extensive than routine quality control tests. The initial test must include those which are contained in the customer and/or supplier product specifications and/or contracts. Results of these tests must be documented.
- *4.4.8.H.3 Frequency of Testing.* The supplier must establish and document the frequency of tests and periodic retests.
- *4.4.9.C.1 Change Management Process.* The supplier must establish and maintain a process to ensure that all requirements and design changes, which may arise at any time during the product life cycle, are managed in a systematic and timely manner and do not adversely affect quality and reliability.
- *4.4.9.C.2 Informing Customers.* The supplier must establish and maintain documented procedures to ensure that customers are informed when design changes affect contractual commitments.
- *4.4.9.H.1 Tracking of Changes.* The supplier must track design changes and must use the results to ensure that the product still fulfills its design intent.
- *4.4.9.H.2 Component Changes.* The supplier must have adequate documented procedures in place to ensure that material or component substitutions or changes do not adversely affect product quality or performance.
- *4.4.9.V.1 Tool Changes.* The supplier must have documented procedures in place to ensure that substitutions or changes to tools used in performing the service do not adversely affect the service quality.

- *4.5.1.S.1 Control of Customer-Supplied Documents and Data.* The supplier must establish and maintain documented procedures to control all customer-supplied documents and data, such as network architecture, topology, capacity, and database, if these documents and data influence the design, verification, validation, inspection and testing, or servicing of the product.
- *4.6.1.C.1 Purchasing Procedures.* The documented purchasing procedures must include product requirements definition; risk analysis and management; qualification criteria; contract definition; satisfaction of proprietary usage, ownership, warranty, and licensing rights; any planned future support for the product; ongoing supply-base management and monitoring; subcontractor selection criteria; subcontractor reevaluation; and feedback to key subcontractors based on data analysis of subcontractor performance.
- *4.8.H.1 Traceability for Recall.* Field replaceable units (FRU) must be traceable throughout the product life cycle in a way that helps suppliers and customers identify products being recalled or needing to be replaced or modified.
- *4.8.H.2 Traceability of Design Changes.* The supplier must establish and maintain documented procedures that provide the ability to trace design changes to identifiable manufacturing dates, lots, or serial numbers.
- *4.8.HS.1 Configuration Management Plan.* The supplier must establish and maintain a configuration management plan, which should include identification and scope of the configuration management activities, a schedule for performing these activities, configuration management tools, configuration management methods and documented procedure(s), organizations and responsibilities assigned to them, level of required control for each configuration item, and point at which items are brought under configuration management.
- *4.8.HS.2 Product Identification.* The supplier must establish and maintain a process for the identification of each product and the level of required control. For each product and its versions, the following must be identified as appropriate: documentation, associated tools needed for product recreation, interfaces to other software and hardware, and software and hardware environment.
- *4.9.H.1 Inspection and Testing.* Inspection and testing results must be recorded and analyzed for the purpose of identifying problem areas.
- *4.9.HV.1 Operational Changes.* Whenever a significant change is made in the established operation, such as a new operator, new

machine, or new technique, a critical examination must be made of the first units or services processed after the change.

- *4.9.HV.2 Operator Qualification.* The supplier must establish operator qualification and requalification requirements for all applicable processes. These requirements, as a minimum, must address employee experience, training, and demonstrated skills. The supplier must communicate this information to all affected employees.
- *4.9.HV.3 Employee Skills List.* The supplier must maintain records of employees and their skills and qualifications to aid in determining work assignments.
- *4.9.S.1 Replication.* The supplier must establish and maintain documented procedures for replication, which should include identification of the master copy and copies to be delivered, the number of copies to be delivered, type of media and associated labeling, identification and packaging of required documentation such as user manuals, and controlling the replication environment to ensure repeatability.
- *4.9.S.2 Release Management.* The supplier must establish and maintain documented procedure(s) to control the release and delivery of software products and documentation.
- *4.9.V.1 Software Used in Service Delivery.* Suppliers must document and implement processes for the maintenance and control of software used in service delivery to ensure continued process capability and integrity.
- *4.9.V.2 Service Delivery Plan.* Suppliers that are responsible for the delivery or implementation of a service, and are not responsible for the design of that service, must comply with the project plan requirements of Element 4.4.2.C.1.
- *4.10.1.HV.1 Inspection and Test Documentation.* Each inspection or testing activity must have detailed documentation. Details should include parameters to be checked with acceptable tolerances; the use of statistical techniques, control charts, and similar methods; a sampling plan, including frequency, sample size, and acceptance criteria; handling of nonconformances; data to be recorded; a defect classification scheme; a method for designating an inspection item or lot; and electrical, functional, and features testing.
- *4.10.1.S.1 Test Documentation.* Software tests must be conducted according to documented procedures and the test plan.
- *4.10.4.H.1 Testing of Repair and Return Products.* Repair and return products must be subjected to the same or equivalent documented final acceptance test procedures as newly manufactured products.

- *4.10.4.H.2 Packaging and Labeling Audit.* The supplier must include a packaging and labeling audit in the quality plan or documented procedures. This may include, for example, marking, labeling, kiting, documentation, customer-specific marking, and correct quantities.
- *4.10.5.HV.1 Inspection and Test Records.* Inspection and test records must include product identification, quantity of product inspected, documented inspection procedures followed, person performing the test or inspection, date of inspection and/or test, and number, type, and severity of defects found.
- *4.11.2.H.1 Identified Equipment.* Inspection, measuring, and test equipment that is either inactive or unsuitable for use must be identified and not used for production. All inspection, measuring, and test equipment that does not require calibration must be identified.
- *4.13.2.C.1 Trend Analysis.* Trend analysis of discrepancies found in nonconforming product must be performed on a defined, regular basis, and results should be used as input for corrective and preventive actions.
- *4.15.1.C.1 Work Areas.* Areas used for handling, storage, and packaging of products must be clean, safe, and organized to ensure that they do not adversely affect quality or personnel performance.
- *4.15.1.C.2 Antistatic Protection.* Antistatic protection must be used where applicable for components and products susceptible to electrostatic discharge (ESD) damage. Components and products to be considered include integrated circuits, printed wiring board assemblies, magnetic tapes and/or disks, and other media used for software or data storage.
- *4.15.2.S.1 Software Virus Protection.* The supplier must establish and maintain methods for software virus prevention, detection, and removal from the deliverable product.
- *4.15.3.H.1 Deterioration.* Where the possibility of deterioration exists, materials in storage must be controlled, such as date stamped and coded, and materials with expired dates must not be used.
- *4.15.6.S.1 Patch Documentation.* The supplier must establish and maintain methods to ensure that all documentation required to describe, test, install, and apply a patch has been verified and delivered with the patch.
- *4.18.C.1 Course Development.* The supplier must establish and maintain a process for planning, developing, and implementing training courses.

- *4.18.C.2 Quality Improvement Concepts.* Those employees who have a direct impact on the quality of the product, including management with executive responsibility, must be trained in the fundamental concepts of quality improvement, problem solving, and customer satisfaction.
- *4.18.C.3 Training Requirements and Awareness.* Training requirements must be defined for all positions that have a direct impact on product quality. Employees must be made aware of training opportunities.
- *4.18.C.4 ESD Training.* All employees with functions that involve any handling, storage, packaging, preservation, or delivery of electrostatic discharge (ESD)–sensitive products must be trained in ESD protection prior to performing their jobs.
- *4.18.C.5 Advanced Quality Training.* Suppliers must offer training in statistical techniques, process capability, statistical sampling, data collection and analysis, problem identification, problem analysis, and corrective and preventive action, as appropriate.
- *4.18.C.6 Training Content.* Where hazardous conditions exist, training content should include task execution, personal safety, awareness of the hazardous environment, and equipment protection.
- *4.19.C.1 Supplier's Support Program.* The supplier's quality program must ensure that customers are provided support to resolve product-related problems.
- *4.19.C.2 Service Resources.* The supplier must provide customer contact employees with appropriate tools, training, and resources necessary to provide effective and timely customer service.
- *4.19.C.3 Notification About Problems.* The supplier must establish and maintain documented procedures to notify all customers who may be affected by a reported service problem.
- *4.19.C.4 Problem Severity.* The customer and supplier must jointly assign severity levels to customer-reported problems based on the impact to the customer. The severity level must be used in determining the timeliness of the supplier's response.
- *4.19.C.5 Problem Escalation.* The supplier must establish and maintain documented escalation procedures to resolve customer-reported problems.
- *4.19.H.1 Supplier's Recall Process.* The supplier must establish and maintain documented procedures to identify and recall products that are unfit to remain in service.

- *4.19.HS.1 Emergency Service.* The supplier must ensure that services and resources are available to support recovery from emergency failures of product in the field throughout its life expectancy.
- *4.19.HS.2 Problem Resolution Configuration Management.* The supplier must establish an interface between problem resolution and configuration management to ensure that fixes to problems are incorporated in future revisions.
- *4.19.HS.3 Installation Plan.* The supplier must establish and maintain a documented installation plan. This plan must identify resources, information, and installation events. Installation events and results must be documented.
- *4.19.S.1 Patching Procedures.* The supplier must establish and maintain documented procedures that guide the decision to solve problems by patching. These documented procedures must address patch development procedures, propagation and resolution, and must be consistent with purchaser needs or contractual requirements for maintenance support. For each patch, the supplier must provide the customer with a statement of impact regarding that patch on the customer's operation.
- *4.19.S.2 Problem Resolution.* The supplier must establish and maintain documented procedures to initiate corrective action once a problem has been diagnosed. These documented procedures should provide guidelines for distinguishing among potential solutions such as patching, immediate source-code corrections, deferring solutions to a planned release, and providing documented "work-around" operational procedures and resolution within a designated timeframe based on the severity of the issue.
- *4.20.1.C.1 Process Measurements.* Process measurements must be developed, documented, and monitored at appropriate points to ensure continued suitability and promote increased effectiveness of processes.
- *4.21 Quality Improvement and Customer Satisfaction*
- *4.21.1 Quality Improvement Program*
- *4.21.1.C.1 Quality Improvement Program.* The supplier must establish and maintain a documented quality improvement program to improve customer satisfaction, the quality and reliability of the product, and other processes, products, and services used within the company.
- *4.21.1.C.2 Employee Participation.* The supplier must have methods for encouraging employee participation in the quality improvement process.

- *4.21.1.C.3 Supplier Performance Feedback.* The supplier must inform employees of its quality performance and the level of customer satisfaction.
- *4.21.2 Customer-Supplier Relationship*
- *4.21.2.C.1 Management Commitment.* Management with executive responsibility must demonstrate active involvement in establishing and maintaining customer-supplier relationships.
- *4.21.2.C.2 Customer-Supplier Communication.* The supplier must establish and maintain documented procedures for communicating with selected customers. These documented procedures must include a strategy and criteria for customer selection, a method for sharing customer and supplier expectations and improving product quality, and a joint review with the customer at defined intervals covering the status of customer-supplier shared expectations, including a method to track issue resolution.
- *4.21.3 Quality Results*
- *4.21.3.C.1 Customer Satisfaction.* The supplier must establish and maintain a method to collect data directly from customers concerning their satisfaction with product provided. The supplier must also collect customer data on how well the supplier meets commitments and its responsiveness to customer feedback and needs. These data must be collected and analyzed, and trends must be kept.
- *4.21.3.H.1 Field Performance Data.* The quality system must include the collection and analysis of field performance data, which can be used to help identify the cause and frequency of equipment failure. In addition, no trouble found (NTF) data must also be maintained. This information must be provided to the appropriate organizations to foster continuous improvement. The quality system must include documented procedures to provide customers with feedback on their complaints in a timely manner.
- *4.21.3.V.1 Service Performance Data.* The quality system must include the collection and analysis of service performance data, which can be used to document the cause and frequency of service failures. This information must be provided to the appropriate organizations to foster continuous improvement of the service.
- *4.21.4 New Product Introduction*
- *4.21.4.C.1 New Product Introduction.* The supplier must establish and maintain documented procedures for introducing new products.

Appendix

A

ISO 9000 Cross-Reference

ISO 9001: 1994 (9004-1 equivalent)	9002	9003	9004-1	9004-2
4.1 Management Responsibility	4.1	4.1	4	5.2
4.2 Quality System (Quality System Elements)	4.2	4.2	5	5.4
4.3 Contract Review (Quality in Marketing)	4.3	4.3	7	6.1
4.4 Design Control (Quality in Specification and Design)	—	—	8	6.2
4.5 Document and Data Control (Quality Documentation and Records)	4.5	4.5	17	5.4.3
4.6 Purchasing (Quality in Procurement)	4.6	—	9	6.2.4.3
4.7 Control of Customer-Supplied Product (Supplier-Provided Equipment to Customers for Service and Service Delivery)	4.7	4.7	—	6.2.4.4
4.8 Product Identification and Traceability (Material Control)	4.8	4.8	11.2	6.2.4.5
4.9 Process Control (Quality of Processes)	4.9	—	10	6.3
4.10 Inspection and Testing (Product Verification)	4.10	4.10	12	6.3.2
4.11 Control of Inspection, Measuring, and Test Equipment (Control of Measuring and Test Equipment)	4.11	4.11	13	—
4.12 Inspection and Test Status (Control of Verification Status)	4.12	4.12	11.7	6.3.4
4.13 Control of Nonconforming Product (Nonconformity)	4.13	4.13	14	6.3.5.2
4.14 Corrective and Preventive Action	4.14	4.14	15	6.3.5
4.15 Handling, Storage, Packaging, Preservation, and Delivery (Postproduction Activities)	4.15	4.15	16	6.2.4.6
4.16 Control of Quality Records	4.16	4.16	17.3	5.4.3
4.17 Internal Quality Audits (Auditing the Quality System)	4.17	4.17	5.4	5.4.4
4.18 Training (Personnel)	4.18	4.18	18	5.3.2.2
4.19 Servicing	4.19	—	16.4	—
4.20 Statistical Techniques (Use of Statistical Methods)	4.20	4.20	20	6.4.3
Benefits, Costs, and Risks	—	—	0.4	—
Product Safety	—	—	19	—

Appendix

B

ISO 9001: 1994 Self-Assessment

Here, arranged in numerical order, are the requirements and guidelines of the 20 elements of ISO 9001: 1994. You can use this section to conduct an informal assessment of the conformance of your facility's quality system to the standard's requirements.

For each requirement, indicate the level of conformance by circling the appropriate score. Enter that score in the column to the right. Total the scores for each section. At the end of this appendix, you can total your scores and evaluate, on a general basis, how your quality system rates against the requirements of ISO 9001: 1994.

ISO 9001: 1994, Element 4.1: Management Responsibility

	Self-assessment				
Components	Strong	Moderate	Weak	N/A	Score
Management has defined and documented a facility quality policy.	10	5	0	10	
Specific quality responsibilities of all employees are defined and documented.	10	5	0	10	
An organization chart exists which shows management structure and relationships.	10	5	0	10	
A specific entity has responsibility for providing sufficient resources and personnel to complete facility tasks.	10	5	0	10	
A management representative, with sufficient authority, has been designated to oversee conformance of quality system to the standard.	10	5	0	10	
Management conducts regular review meetings to assess the quality system and ensure its continued effectiveness. Minutes of such reviews are documented. The location of this documentation is specified.	10	5	0	10	
Procedures covering all activities specified by this element exist.	10	5	0	10	
				TOTAL	

ISO 9001: 1994, Element 4.2: Quality System

	Self-assessment				
Components	Strong	Moderate	Weak	N/A	Score
Your facility has established and implemented a documented quality system.	10	5	0	10	
Management has specified who, within the facility, is responsible for, and holds authority for, the quality system.	10	5	0	10	
Quality activities are consistent with your facility's quality policy and quality system documentation.	10	5	0	10	
Documented procedures exist and reference your facility's quality policy and all 138 requirements of the ISO 9001: 1994 standard.	10	5	0	10	
Quality plans, which may be in the form of documented procedures, have been prepared to control quality system activities.	10	5	0	10	
Those controls, processes, equipment, fixtures, resources, and skills needed to satisfy your facility's quality objectives have been identified and acquired.	10	5	0	10	
The compatibility of designs, processes, and procedures for installation, servicing, and inspection and test activities have been verified.	10	5	0	10	
Verification activities are suitable for the product and the process.	10	5	0	10	
Standards of acceptability have been clarified.	10	5	0	10	
Quality records are maintained.	10	5	0	10	
				TOTAL	

ISO 9001: 1994, Element 4.3: Contract Review

Components	Self-assessment				Score
	Strong	Moderate	Weak	N/A	
Facility has defined responsibility and authority for contract review activities.	10	5	0	10	
Facility's contract review system ensures that customer requirements are adequately defined and documented.	10	5	0	10	
Facility's contract review system ensures that inconsistencies between customer requirements and facility standards are detected and resolved.	10	5	0	10	
Facility's contract review system ensures that facility has full capability to meet its obligations under the contract.	10	5	0	10	
Responsibility and authority for each step of facility's contract review system is specified by title or function.	10	5	0	10	
Records of contract review activities are maintained.	10	5	0	10	
Detailed procedures covering the systems herein are available.	10	5	0	10	
				TOTAL	

ISO 9001: 1994, Element 4.4: Design Control

Components	Self-assessment				Score
	Strong	Moderate	Weak	N/A	
Facility has defined responsibility and authority for the design process. References to documents listing qualifications are included.	10	5	0	10	
Facility has a system that defines the design planning and development process.	10	5	0	10	
Facility possesses a document that displays the relationships and interfaces among functions involved in the design process, as well as interfaces between the design process and other process elements.	10	5	0	10	
Sources of design input are identified. Facility creates a particular document, resulting from input sources, that crystallizes design requirements.	10	5	0	10	
Facility has defined its types of design output, such as blueprints and design checklists, and can present samples of these.	10	5	0	10	
The facility's system for verifying that designs meet input requirements is defined. The system includes at least two of the following methods: design reviews, qualification tests, alternative calculations, or comparison with proven designs.	10	5	0	10	
The facility's design review system takes into account various causes of design changes. The facility has a procedure for addressing, evaluating, and implementing needed changes.	10	5	0	10	
Detailed procedures covering the systems herein are available.	10	5	0	10	
				TOTAL	

ISO 9001: 1994, Element 4.5: Document and Data Control

Components	Self-assessment				Score
	Strong	Moderate	Weak	N/A	
Facility has defined responsibility and authority for the creation, distribution, revision, and control of quality-related documents and data.	10	5	0	10	
Types of quality-related documents and data in use at the facility have been identified.	10	5	0	10	
A procedure exists that specifies how quality-related documents and data are created and stored for employees needing access to them.	10	5	0	10	
A procedure exists that governs how quality-related documents and data are modified and approved and which specifies the ways in which obsolete editions are withdrawn and discarded. It also specifies a number of changes that can be made to a document or data before a complete reissue is required.	10	5	0	10	
A list of the current editions of all quality-related documents and data is maintained.	10	5	0	10	
Detailed procedures covering the systems herein are available.	10	5	0	10	
				TOTAL	

ISO 9001: 1994, Element 4.6: Purchasing

	Self-assessment				
Components	Strong	Moderate	Weak	N/A	Score
Facility has defined responsibility and authority for (a) purchasing activities and (b) quality of purchased products and services.	10	5	0	10	
Facility has a documented system for selecting subcontractors that ensures that authorized subcontractors can meet specified requirements.	10	5	0	10	
Facility has a procedure that specifies how purchasing data are to be communicated to subcontractors in a way that precludes ambiguity or confusion.	10	5	0	10	
Facility has a procedure for verifying conformity of purchased products and services either at facility or at source. Such verification does not absolve subcontractors of responsibility of meeting specified requirements.	10	5	0	10	
Detailed procedures covering the systems herein are available.	10	5	0	10	
				TOTAL	

ISO 9001: 1994, Element 4.7: Control of Customer-Supplied Product

	Self-assessment				
Components	Strong	Moderate	Weak	N/A	Score
Facility has defined responsibility and authority for the scheduling, handling, and storage of customer-supplied items.	10	5	0	10	
Procedures exist for verification of incoming customer-supplied product to determine conformance with respect to features, quantity, and condition.	10	5	0	10	
Facility has a system for safeguarding customer-supplied product.	10	5	0	10	
Facility has a program of regular communication with owners of customer-supplied product to resolve any nonconformances.	10	5	0	10	
Detailed procedures covering the systems herein are available.	10	5	0	10	
				TOTAL	

ISO 9001: 1994, Element 4.8: Product Identification and Traceability

Components	Self-assessment				Score
	Strong	Moderate	Weak	N/A	
Facility has defined responsibility and authority for assessing the need for product identification and traceability, if deemed appropriate. Facility has defined responsibility for managing these activities.	10	5	0	10	
Facility has defined method for determining if, and where, traceability is required.	10	5	0	10	
If appropriate and/or required by customer contract, procedures exist for identifying products via tagging or other means or services via accompanying documentation.	10	5	0	10	
If appropriate and/or required by customer contract or legal mandate, procedures exist for tracing the origin, application, and/or location of products or services.	10	5	0	10	
Detailed procedures covering the systems herein are available.	10	5	0	10	
				TOTAL	

ISO 9001: 1994, Element 4.9: Process Control

Components	Self-assessment				Score
	Strong	Moderate	Weak	N/A	
Facility has defined responsibility and authority for process control activities.	10	5	0	10	
Process areas having an impact on quality have been clearly defined.	10	5	0	10	
Procedures governing the rigorous measurement and monitoring of special processes exist and are available to all affected personnel.	10	5	0	10	
Records for qualified processes, equipment and personnel are maintained, as appropriate.	10	5	0	10	
				TOTAL	

ISO 9001: 1994, Element 4.10: Inspection and Testing

	Self-assessment				
Components	Strong	Moderate	Weak	N/A	Score
Facility has defined responsibility and authority for receiving, in-process, and final inspection and testing systems.	10	5	0	10	
Facility's system provides for release of supplied product subject to recall. System includes mechanism for identification, traceability, control, and recall of such products, under defined responsibility and authority.	10	5	0	10	
Facility has system that ensures early recognition of nonconforming product at various vital in-process stages and provides for identification and disposition of such materials.	10	5	0	10	
Facility has system that ensures that output meets specified characteristics prior to release. System includes means for clear identification of conforming versus nonconforming output.	10	5	0	10	
Detailed procedures covering the systems herein are available.	10	5	0	10	
				TOTAL	

ISO 9001: 1994, Element 4.11: Control of Inspection, Measuring, and Test Equipment

Components	Self-assessment				
	Strong	Moderate	Weak	N/A	Score
Facility has defined responsibility and authority for inspection, measuring, and test equipment.	10	5	0	10	
Facility has procedure for selecting measurements, determining the accuracy required, and acquiring equipment that meets these requirements.	10	5	0	10	
Facility has system for verifying equipment needed and measurement accuracy and for confirming that environmental conditions permit reliable use.	10	5	0	10	
Facility has system for calibration and adjustment of equipment at prescribed intervals against (1) nationally recognized standards or (2) documented benchmark where no nationally recognized standards exist.	10	5	0	10	
Facility maintains full documentation of calibration procedures and results.	10	5	0	10	
Facility identifies all equipment with indicator of calibration status.	10	5	0	10	
Facility has system for reverifying previous inspection, measuring, and test results when equipment is found to be out of calibration.	10	5	0	10	
Facility has system for maintaining and storing equipment that ensures that accuracy and fitness for use are preserved. This system includes means of protecting equipment from unauthorized adjustment.	10	5	0	10	
Detailed procedures covering the systems herein are available.	10	5	0	10	
				TOTAL	

ISO 9001: 1994, Element 4.12: Inspection and Test Status

	Self-assessment				
Components	Strong	Moderate	Weak	N/A	Score
Facility has defined responsibility and authority for identifying inspection and test status of raw materials, supplied items, work in progress, and finished output.	10	5	0	10	
Facility has system for showing, at all stages, whether such items have not been inspected, been inspected and accepted, been inspected and on hold awaiting resolution, or been inspected and rejected.	10	5	0	10	
Detailed procedures covering the systems herein are available.	10	5	0	10	
				TOTAL	

ISO 9001: 1994, Element 4.13: Control of Nonconforming Product

	Self-assessment				
Components	Strong	Moderate	Weak	N/A	Score
Facility has defined responsibility and authority for identifying nonconforming product and evaluating, segregating, and disposing of it.	10	5	0	10	
Procedures exist for identifying non-conforming product and evaluating it, including identifying its source.	10	5	0	10	
Procedures exist for segregating the product to prevent inadvertent use.	10	5	0	10	
Procedures exist governing methods of disposal, including rework or repair to specified requirements, acceptance with or without repair by concession, modification for other use, or scrapping.	10	5	0	10	
System includes specified means of documenting all of the above.	10	5	0	10	
Detailed procedures covering the systems herein are available.	10	5	0	10	
				TOTAL	

ISO 9001: 1994, Element 4.14: Corrective and Preventive Actions

Components	Self-assessment				Score
	Strong	Moderate	Weak	N/A	
Facility has defined responsibility and authority for designing, implementing, and documenting corrective actions.	10	5	0	10	
Facility has procedures for detecting causes of nonconformities, initiating corrective actions, controlling their implementation, verifying their effectiveness, and documenting procedural changes in order to prevent recurrence.	10	5	0	10	
Detailed procedures covering the systems herein are available.	10	5	0	10	
				TOTAL	

ISO 9001: 1994, Element 4.15: Handling, Storage, Packaging, Preservation, and Delivery

Components	Self-assessment				Score
	Strong	Moderate	Weak	N/A	
Facility has defined responsibility and authority for maintaining the quality of all materials during handling, storage, packaging, preservation, and delivery.	10	5	0	10	
Procedures exist for handling of materials to prevent damage or deterioration.	10	5	0	10	
Procedures exist that ensure security of storage areas to avert environmental or human damage, deterioration, or shrinkage.	10	5	0	10	
Procedures exist governing audit and assessment methods and means of documentation.	10	5	0	10	
Procedures specify handling methods to preserve materials in a state that conforms with specified requirements and accurate marking processes to avert mishandling.	10	5	0	10	
Detailed procedures covering the systems herein are available.	10	5	0	10	
				TOTAL	

ISO 9001: 1994, Element 4.16: Control of Quality Records

	Self-assessment				
Components	Strong	Moderate	Weak	N/A	Score
Facility has defined responsibility and authority for the creation, maintenance, retention, and systematic disposal of quality records.	10	5	0	10	
Records exist that demonstrate effective operation of the quality system.	10	5	0	10	
Records document achievement of the required quality levels and remedial actions taken in response to nonconformances.	10	5	0	10	
Records are readily retrievable by all authorized to use them.	10	5	0	10	
Records are discarded on an orderly and systematic basis when documented retention intervals have passed.	10	5	0	10	
Detailed procedures covering the systems herein are available.	10	5	0	10	
				TOTAL	

ISO 9001: 1994, Element 4.17: Internal Quality Audits

Components	Self-assessment				
	Strong	Moderate	Weak	N/A	Score
Facility has defined responsibility and authority for planning, scheduling, conducting, documenting, and benefiting from internal quality audits.	10	5	0	10	
Procedures place emphasis on auditing areas of critical importance to the quality system or which have posted a history of nonconformances or other problems.	10	5	0	10	
Procedures specify the qualifications of the personnel assigned to perform internal audits.	10	5	0	10	
Procedures exist governing the conduct of internal audits, including safeguards against conflicts of interest and provision for follow-up actions.	10	5	0	10	
Results of internal audits are brought to the attention of the personnel responsible for the area audited.	10	5	0	10	
Management of audited areas is held accountable for taking corrective actions on or before specified dates.	10	5	0	10	
Detailed procedures covering the systems herein are available.	10	5	0	10	
				TOTAL	

ISO 9001: 1994, Element 4.18: Training

Components	Self-assessment				Score
	Strong	Moderate	Weak	N/A	
Facility has defined responsibility and authority for setting quality-related employment qualifications, assessing needs, providing training, and maintaining records.	10	5	0	10	
Qualifications, in terms of education, training, and experience, exist for each position that affects quality.	10	5	0	10	
Procedures exist that provide for evaluating training needs for all quality-sensitive positions periodically.	10	5	0	10	
Records of all training activities, both individual and corporate, are maintained.	10	5	0	10	
Detailed procedures covering the systems herein are available.	10	5	0	10	
				TOTAL	

ISO 9001: 1994, Element 4.19: Servicing

Components	Self-assessment				Score
	Strong	Moderate	Weak	N/A	
Facility has defined responsibility and authority for servicing activities.	10	5	0	10	
Facility has established and maintained procedures for performing service.	10	5	0	10	
Facility has procedures for verifying and reporting that servicing meets established customer requirements.	10	5	0	10	
Detailed procedures covering the systems herein are available.	10	5	0	10	
				TOTAL	

ISO 9001: 1994, Element 4.20: Statistical Techniques

	Self-assessment				
Components	Strong	Moderate	Weak	N/A	Score
Facility has defined responsibility and authority for evaluating the need for statistical techniques throughout the facility and process.	10	5	0	10	
Facility has procedures for assessing statistical techniques to verify critical product or service characteristics, assess process capability, and other purposes.	10	5	0	10	
Detailed procedures covering the systems herein are available.	10	5	0	10	
				TOTAL	

Your ISO 9001: 1994 Self-Assessment

To get a snapshot look at how ready your quality system is for ISO 9001: 1994 registration, add up the scores from each of the 20 elements. Enter the total in the space below, and plot your score on the chart.

Score ———

1140	847	565	283

848–1140: *Strong* — Most elements of your system conform to the basic ISO 9001: 1994 quality system standard. Examine the areas with weak scores, and implement enhancements. Refer to ISO 9004-1 for additional guidance. It is also advisable to arrange for a preassessment from an ISO 9000 registrar.

566–847: *Moderate* — Many elements of your system conform to the basic ISO 9001: 1994 quality system standard, but a significant part of your quality system either does not conform to the standard or is not present. Pinpoint these areas, and implement the systems required. For guidance, you can refer to official manuals such as ISO 9004-1 or acquire the services of an ISO 9000 consulting and training specialist.

284–565: *Weak* — Much of your quality system does not conform to the basic ISO 9001: 1994 quality system standard. A great deal of orientation and training in the specifics of the standard are necessary. If your facility is serious about ISO 9001 registration, establish a steering committee and acquire the services of an ISO 9000 consulting and training specialist.

0–283: *Poor* — Your quality system is practically nonexistent. It is imperative that your entire process undergo review. We recommend a thorough regimen of training in total quality management tools and techniques, as well as ISO 9000 training and consultations, before investing time and effort in ISO 9000 registration.

Appendix

C

Sample Quality Manual

XYZ Typesetting Co. Quality Manual	Section 0.0 TABLE OF CONTENTS	Page 1 of 1

0.0 Table of Contents

CONTROLLED CIRCULATION
Copy 11 Director of Quality

Prepared By Approved By	Jane Hopkins Damon Keith	Issue Number Issue Date	1 Feb. 6, 1995

XYZ Typesetting Co. Quality Manual	Section 0.1 QUALITY POLICY STATEMENT	Page 1 of 1

0.1 Quality Policy Statement

The XYZ Typesetting Co. is totally committed to understanding and meeting the quality needs and expectations of all of our customers.

I personally affirm this commitment and have established a comprehensive quality management system that will allow our company to meet all the requirements of the ISO 9001: 1994 quality system standard. We are committing ourselves to a strategy of continuous improvement, relentlessly seeking to learn the expectations of our customers, and striving to meet and exceed those expectations at every juncture. We intend for these efforts to help us meet the following goals:

- Double our market share within 5 years
- Maintain our position at the cutting edge of typesetting and publishing technology
- Facilitate continuous improvement in all aspects of our quality management system
- Achieve outstanding financial performance as measured by return on investment and return on assets
- Maintain our reputation for delivering quality products, service and civic responsibility

The entire XYZ Typesetting team will adhere to the spirit and intent of the firm's quality policy, as well as the directives of this quality assurance manual and its supporting quality system documentation. We will continue to aggressively strive to ensure that customer satisfaction is achieved at all times, and in all things.

Signed:

Harry S. Lewis
President February 1995

Prepared By Approved By	Cass Timberlane Carol Kennicott	Issue Number Issue Date	1 Feb. 6, 1995

XYZ Typesetting Co. Quality Manual	Section 0.2 COMPANY BACKGROUND	Page 1 of 1

0.2 Company Background

The XYZ Typesetting Co. was founded by Harry S. Lewis and George F. Babbitt in Sauk Centre, Minn. in 1968. Since its establishment, the company has grown from a single-person entrepreneurship to a 25,000-square-foot plant employing the latest in computer-driven typesetting systems. XYZ now has a printing division, a direct-mail division, and a telemarketing services division.

XYZ Typesetting has major clients in the advertising and printing industries and maintains a worldwide market for its products and technical support services. The company was built on the idea that its customers are the most important part of the business. It has grown by providing quality products and innovative ideas and by using the leading edge of technology concepts and methods.

Prepared By	Max Gottlieb	Issue Number	1
Approved By	Leora Tozer	Issue Date	Feb. 6, 1995

XYZ Typesetting Co. Quality Manual	Section 0.3 AMENDMENT RECORD	Page 1 of 1

0.3 Amendment Records

This quality assurance manual (QAM) contains only the pages issued by this facility. The management representative (MR) will process all authorized changes, inserting amended pages into the official distribution copies. The MR will see that all down-level and/or obsolete pages are withdrawn from use and disposed of to prevent unintentional usage.

This QAM is a controlled-copy document. The master copy (MC) of this QAM is maintained by the MR. This MC shall be used as the final authority, regarding the latest revision level and amendment status for the XYZ Typesetting Co. QAM.

Date	Section/page	Details	Approval
1/5/94	All	Complete revision of QAM	HLS

Prepared By Approved By	T. Earl Petty Reginald K. Dwight	Issue Number Issue Date	1 Feb. 6, 1995

XYZ Typesetting Co. Quality Manual	Section 0.4 CONTROLLED CIRCULATION LIST	Page 1 of 1

0.4 Controlled Circulation List

Copy no.	Copy custodian
1	President
2	Senior Vice President
3	Vice President of Operations
4	Vice President of Direct Marketing
5	Vice President of Telemarketing
6	Vice President of Client Services
7	Director of Administration
8	Director of Finance
9	Director of Planning
10	Director of Marketing
11	Director of Quality
12	Director of Human Resources
13	Director of Purchasing
14	Registrar

Prepared By	Walter Smilanovich	Issue Number	1
Approved By	Homer Winslow	Issue Date	Feb. 6, 1995

XYZ Typesetting Co. Quality Manual	Section 0.5 GLOSSARY	Page 1 of 1

0.5 Glossary

Quality Assurance Manual Glossary

MR—Management Representative
QAM—Quality Assurance Manual
MC—Master Copy
XYZ—XYZ Typesetting Co.
Standard(s)—Industry, national, and international quality standard ISO 9001: 1994
R&A—Responsibility and Authority
C&QP—Control and Quality Plan

Prepared By Approved By	William Laxton Augie Ruiz	Issue Number Issue Date	1 Feb. 6, 1995

XYZ Typesetting Co. Quality Manual	Section 1.0 MANAGEMENT RESPONSIBILITY	Page 1 of 2

1.0 Management Responsibility

1.1 Scope and purpose

The quality system described in this section of the QAM conforms to the requirements of the standard(s): Element 4.1, Management Responsibility

1.2 Responsibility and authority (R&A)

The R&A for overall administration of quality activities are shared by the president and the MR. The associates of XYZ have the responsibility to carry out all quality activities in support of its quality policy, quality system documentation, and customer requirements. Each associate has been granted authority in order to meet specified requirements.

1.3 Quality system requirements

1.3.1 Quality policy. A company quality policy has been established identifying quality goals and objectives. This policy has been communicated to all employees and is maintained as the highest priority within the company. Each associate understands his or her role.

1.2 Responsibility and authority. The R&A for all activities that affect quality have been defined in both quality system documentation and job descriptions.

1.3.3 Resources. Adequate resources required to complete quality system activities have been defined in both quality system documentation and job descriptions.

1.3.4 Management representative. The MR has been appointed by the company president. The MR has been granted full authority to establish, implement, maintain, and report on quality management system activities.

Prepared By Approved By	Declan McManus Paul B. Hewson	Issue Number Issue Date	1 Feb. 6, 1995

XYZ Typesetting Co. Quality Manual	Section 1.0 MANAGEMENT RESPONSIBILITY	Page 2 of 2

1.3.5 Management review. The MR carries out scheduled management review meetings with executive management. These reviews determine the effectiveness and suitability of the implemented quality system requirements. Minutes of these review meetings are maintained.

1.4 Related and support documentation

2QP01-0001 Corporate Job Description(s)
2QP01-0002 Corporate Resource Assignment
2QP01-0003 Management Review Meetings Procedure

Prepared By Approved By	Declan McManus Paul B. Hewson	Issue Number Issue Date	1 Feb. 6, 1995

XYZ Typesetting Co. Quality Manual	Section 2.0 QUALITY SYSTEM	Page 1 of 1

2.0 Quality System

2.1 Scope and purpose

The quality system described in this section of the QAM conforms to the requirements of the standard(s): Element 4.2, Quality System

2.2 Responsibility and authority (R&A)

The R&A for carrying out quality system activities have been assigned to the director of quality and the MR. All associates have the responsibility to carry out their work assignments in accordance with the quality policy and quality system documentation. The associates have been granted authority to complete the activities assigned in order to meet specified requirements.

2.3 Quality system requirements

2.3.1 Quality system manual and procedures. A quality system manual and procedures have been created to address all requirements of the standard.

2.3.2 Quality planning. Quality planning activities are carried out to ensure that all specified requirements have been addressed and met. C&QP and quality system documentation control the processes and methods used to meet these requirements. Quality planning methods and practices identify and control the following:

All quality-related activities are governed by procedures and written instructions. Within the quality system emphasis is placed on the following:

- Acquisition of equipment
- Compatibility of design, production process, installation, servicing, inspection, and test procedures
- Measurement requirements
- Suitability of verification activities
- Standards of acceptability
- Preparation of quality records

2.4 Related and supporting documentation

2QP02-0001 Quality System Planning Procedures
2QP02-0002 Control and Quality Planning Documents

Prepared By Approved By	Ed Husserl W.F. Godot	Issue Number Issue Date	1 Feb. 6, 1995

XYZ Typesetting Co. Quality Manual	Section 3.0 CONTRACT REVIEW	Page 1 of 1

3.0 Contract Review

3.1 Scope and purpose

The quality system described in this section of the QAM conforms to the requirements of the standard(s): Element 4.3, Contract Review

3.2 Responsibility and authority (R&A)

The R&A for carrying out quality system activities have been assigned to the director of marketing. All associates have the responsibility to carry out their work assignments in accordance with the quality policy and quality system documentation. The associates have been granted authority to complete the activities assigned in order to meet specified requirements.

3.3 Quality system requirements

3.3.1 Contract review and amendments. Procedures exist to control the methods and practices used to complete customer contract reviews. All contracts, verbal and written, adequately define the specified requirements. Differences between specified requirements and capabilities will be resolved. Amendments to contracts will be defined and communicated to all affected functional groups.

3.3.2 Records. Records of contract reviews are maintained.

3.4 Related and support documentation

2QP03-0001 Contract Review Procedures
2QP03-0002 Contract Amendment Procedures
2QP03-0003 Contract—General Requirements List

Prepared By Approved By	Martin Kingsblood Doremus Walker	Issue Number Issue Date	1 Feb. 6, 1995

4.0 Design Control

4.1 Scope and purpose

The quality system described in this section of the QAM conforms to the requirements of the standard(s): Element 4.4, Design Control

4.2 Responsibility and authority (R&A)

The R&A for carrying out quality system activities have been assigned to the director of engineering and the MR. All associates have the responsibility to carry out their work assignments in accordance with the quality policy and quality system documentation. The associates have been granted authority to complete the activities assigned in order to meet specified requirements.

4.3 Quality system requirements

Documented procedures exist to control all the following quality system activities and requirements.

4.3.1 Design and development planning. Plans outlining design activities and their associated schedules are developed for each design project. These plans define various R&A and are used to ensure that engineering personnel, with appropriate skills, are assigned to each design project. All plans are formally documented and are modified as the design activities are evolving.

4.3.2 Organizational and technical interfaces. All functional groups are involved in reviewing and evaluating the aspects of the design through its various stages. The assigned project engineering group has the R&A for fulfilling these activities.

4.3.3 Design input. The engineering group interfaces with the marketing function in order to ensure a complete understanding of the customer's requirements and other requirements, such as those of industry and government.

XYZ Typesetting Co. Quality Manual	Section 4.0 DESIGN CONTROL	Page 2 of 2

4.3.4 Design output. Engineering analysis, evaluation, and other suitable methods are used to verify and validate that design output(s) have addressed and met all specified requirements. These activities are completed and the results are reviewed by engineering and support function personnel prior to release. The results of each design analysis or evaluation will address each specific customer requirement and indicate that criteria have been met in an acceptable manner.

4.3.5 Design review. At various stages of the design, formal structured and documented design review activities are held. These reviews are carried out to determine if all specified requirements have been addressed and all functional support organizations are cognizant of the design phase and essential information.

4.3.6 Design verification and validation (see 4.3.3, Design input, and 4.3.4, Design output). Design verification and validation activities are performed to ensure that the design output meets design input requirements. Design verification measures are recorded. Design verification may include prototyping, using the same subcontractor(s), equipment, tooling, and manufacturing processes planned for use in production, and engineering performance testing to determine design reliability, durability, and fitness for use. These tests will be performed based on program schedules and other needs.

4.3.7 Design changes. Design modifications are identified (red-line marked), documented, reviewed (see 4.3.5, Design review), and approved by the various functional organizations affected by the associated changes. These activities are carried out prior to design change implementation. These controls apply to all changes, regardless of the origin or ownership of the design.

4.4 Related and support documentation

2QP04-0001 Design Control Procedure
2QP04-0002 Engineering/Marketing Input Review Methods & Practices
2QP04-0003 Design Verification and Validation Methods & Practices
2QP04-0004 Design Review and Change Procedures

Prepared By Approved By	John Sloane Maury Prendergast	Issue Number Issue Date	1 Feb. 6, 1995

XYZ Typesetting Co. Quality Manual	Section 5.0 DOCUMENT AND DATA CONTROL	Page 1 of 1

5.0 Document and Data Control

5.1 Scope and purpose

The quality system described in this section of the QAM conforms to the requirements of the standard(s): Element 4.5, Document and Data Control.

5.2 Responsibility and authority (R&A)

The R&A for carrying out quality system activities have been assigned to the director of quality and the MR. All associates have the responsibility to carry out their work assignments in accordance with the quality policy and quality system documentation. The associates have been granted authority to complete the activities assigned in order to meet specified requirements.

5.3 Quality system requirements

Documented procedures controlling all aspects of the creation, review, approval, modification, issue, release, identification of special characteristics, and other activities associated with document and data control are adhered to. These controls apply to all documents, regardless of their origin.

Current revision levels of all documents are maintained in the areas where the work described in the documents is carried out. An on-line listing of all quality-related documentation is maintained. This listing includes current revision-level information and is available to all associates that need this information to carry out their activities.

Down-level and obsolete documents are marked to ensure that they are not used to make decisions that affect quality. These documents are stamped by a member of the document control organization. Historical data are maintained for reference purposes.

All customer-originated documents are received by the marketing group and forwarded to the document control organization for review and distribution to the affected functional groups. Implementation into production practice is carried out and documented to indicate the effective date of the implementation.

5.4 Related and support documentation

2QP05-0001 Document Control Procedure
2QP05-0002 Document Modification Procedure
2QP05-0003 Document Marking Practices
2QP05-0004 Document Revision Level On-line Use

Prepared By	Ben Scribner	Issue Number	3
Approved By	Damon Keith	Issue Date	7/2/93

XYZ Typesetting Co. Quality Manual	Section 6.0 PURCHASING	Page 1 of 1

6.0 Purchasing

6.1 Scope and purpose

The quality system described in this section of the QAM conforms to the requirements of the standard(s): Element 4.6, Purchasing.

6.2 Responsibility and authority (R&A)

The R&A for carrying out quality system activities have been assigned to the director of quality and the MR. All associates have the responsibility to carry out their work assignments in accordance with the quality policy and quality system documentation. The associates have been granted authority to complete the activities assigned in order to meet specified requirements.

6.3 Quality system requirements

Documented procedures controlling all aspects of purchasing are in use. These controls apply to raw materials, components, assemblies, subassemblies, final assemblies, services, and other materials. These procedures control the following activities:

- Identification and selection of vendor(s)
- Evaluation of potential and current vendor(s)
- Determination of vendor(s) capabilities and abilities to meet specified requirements
- Determination of the controls that will be applied to each vendor
- Development of vendor capabilities to meet standard(s) requirements
- Regular review of vendor performance against quality, cost, and delivery requirements
- All purchase order (PO) information contains adequate detail to ensure that all specified requirements have been adequately defined
- Reviews of all PO and related information will be carried out by purchasing personnel in advance of the PO being placed with the vendor

6.4 Related and support documentation

2QP06-0001 Procurement—Raw Material—Procedure
2QP06-0002 Procurement—Components/Assemblies—Procedure
2QP06-0003 Procurement—Capital Equipment—Procedures
2QP06-0004 Procurement—Vendor Selection & Control Procedure

Prepared By	Gary Grafton	Issue Number	1
Approved By	Ike Blessitt	Issue Date	Feb. 6, 1995

XYZ Typesetting Co. Quality Manual	Section 7.0 CONTROL OF CUSTOMER-SUPPLIED PRODUCT	Page 1 of 1

7.0 Control of Customer-Supplied Product

7.1 Scope and purpose

The quality system described in this section of the QAM conforms to the requirements of the standard(s): Element 4.7, Control of Customer-Supplied Product.

7.2 Responsibility and authority (R&A)

The R&A for carrying out quality system activities have been assigned to the director of quality, director of manufacturing, and the MR. All associates have the responsibility to carry out their work assignments in accordance with the quality policy and quality system documentation. The associates have been granted authority to complete the activities assigned in order to meet specified requirements.

7.3 Quality system requirements

Documented procedures controlling all aspects of handling customer-supplied product or services are in use. Customer-supplied product includes

- Tooling and equipment
- Inspection and test equipment
- Raw materials, components, and assemblies
- Packaging materials
- Engineering samples and models
- Specification drawings and other reference documents

These procedures control the following activities:

- Inspection or similar verification activities to determine acceptability
- Storage and handling to protect from loss, damage, or deterioration
- Maintenance and calibration to ensure acceptability for use
- Regular reports indicating the status of customer-supplied product

7.4 Related and support documentation

2QP07-0001 Customer Products/Services Procedure
2QP07-0002 Customer Reporting Procedure

Prepared By Approved By	Gideon Planish Elmer F. Gantry	Issue Number Issue Date	1 Feb. 6, 1995

XYZ Typesetting Co. Quality Manual	Section 8.0 PRODUCT IDENTIFICATION AND TRACEABILITY	Page 1 of 1

8.0 Product Identification and Traceability

8.1 Scope and purpose

The quality system described in this section of the QAM conforms to the requirements of the standard(s): Element 4.8, Product Identification and Traceability.

8.2 Responsibility and authority (R&A)

The R&A for carrying out quality system activities have been assigned to the director of quality, director of manufacturing, and the MR. All associates have the responsibility to carry out their work assignments in accordance with the quality policy and quality system documentation. The associates have been granted authority to complete the activities assigned in order to meet specified requirements.

8.3 Quality system requirements

Documented procedures exist to control all aspects of product identification and traceability. These controls apply to all situations where specified requirements indicate a need for identification and/or traceability. These controls include the following:

- Raw materials, components, subassemblies, and final assemblies are identified through the use of specific packaging and labeling, assignment of part numbers, engineering change or revision levels, or other methods.
- Part number(s) and revision-level identification may be a permanent or temporary part of the product. This identification is maintained from the time raw materials or components are received into the system until(as required) delivery to the end-user/customer.
- Traceability is maintained where it has been identified as a specified requirement. This traceability is documented to permit control of these products or materials as conditions may warrant.

8.4 Related and support documentation

2QP08-0001 Product Identification and Traceability Procedure
2QP08-0002 Product Traceability and Control Procedure
2QP08-0003 Product Identification Methods and Options
2QP08-0004 Product Traceability Records Maintenance

Prepared By Approved By	Frank Shallard Lori Wolfenschmidt	Issue Number Issue Date	1 Feb. 6, 1995

XYZ Typesetting Co. Quality Manual	Section 9.0 PROCESS CONTROL	Page 1 of 2

9.0 Process Control

9.1 Scope and purpose

The quality system described in this section of the QAM conforms to the requirements of the standard(s): Element 4.9, Process Control.

9.2 Responsibility and authority (R&A)

The R&A for carrying out quality system activities have been assigned to the director of quality, director of manufacturing, and the MR. All associates have the responsibility to carry out their work assignments in accordance with the quality policy and quality system documentation. The associates have been granted authority to complete the activities assigned in order to meet specified requirements.

9.3 Quality system requirements

Documented controls, plans, and procedures exist to govern the methods and practices used to complete manufacturing activities and processes. The controls include

- Documented procedures
- Instructions
- Process sheets
- Manufacturing routings
- Quality plans exist to direct the methods and practices used to complete manufacturing operations, and these methods and practices are employed.
- Manufacturing equipment is selected based on needs and the ability to satisfy specified requirements.
- Manufacturing process parameters are identified and monitored to ensure proper control.
- Process methods, practices, and operational personnel are reviewed, approved, and qualified.
- Workmanship criteria are identified and documented in the clearest practical manner to ensure conformance with specified requirements.
- Process equipment and associated tooling are maintained to ensure continuing process capabilities.
- New orders (jobs) use specific setup instructions to ensure that these setup activities are completed in a uniform and consistent manner.
- Records are maintained.

Prepared By Approved By	Yvonne Turgenev Soames Forsythe	Issue Number Issue Date	1 Feb. 6, 1995

Where the results of manufacturing processes, such as heat treating, painting, and welding, cannot be fully verified by subsequent inspection and testing of the product, these processes are carried out by qualified operators and require continuous monitoring and control of process parameters to ensure that the specified requirements are met. Records are maintained for qualified processes, equipment, and personnel, as appropriate

9.4 Related and support documentation

2QP09-0001 Manufacturing Process Control Procedure
2QP09-0002 Preventive and Predictive Maintenance Procedure
2QP09-0003 Preliminary and Ongoing Process Capability Practices
2QP09-0004 Job Setup Practices

XYZ Typesetting Co. Quality Manual	Section 10.0 INSPECTION AND TESTING	Page 1 of 1

10.0 Inspection and Testing

10.1 Scope and purpose

The quality system described in this section of the QAM conforms to the requirements of the standard(s): Element 4.10, Inspection and Testing.

10.2 Responsibility and authority (R&A)

The R&A for carrying out quality system activities have been assigned to the director of quality and the MR. All associates have the responsibility to carry out their work assignments in accordance with the quality policy and quality system documentation. The associates have been granted authority to complete the activities assigned in order to meet specified requirements.

10.3 Quality system requirements

Documented procedures controlling all phases of receiving, in-process, and final/finished goods inspection and testing (I&T) are in use. These procedures and/or quality plans control the following activities:

- No received materials or products are used until all documented I&T has been completed.
- Urgent release of received or in-process products is not a current part of our quality system practices (see 4.6, Purchasing).
- The type and extent of vendor controls used are determined by the amount and type of receiving I&T activities.
- In-process I&T is carried out in accordance with documented procedures and quality plans.
- No products are used or advanced to their next process step until all required I&T activities have been completed.
- I&T is carried out in accordance with documented procedures and quality plans for final/finished goods.
- No final/finished goods are dispatched to the end-user or customer until all required I&T has been completed.
- Records of I&T results are maintained.
- Documentation exists to indicate who has authority to permit the release of product after completion of I&T.

10.4 Related and support documentation

2QP10-0001 Receiving Inspection and Testing Procedure
2QP10-0002 In-process Inspection and Testing Procedure
2QP10-0003 Final/Finished Goods Inspection and Testing Procedures

Prepared By Approved By	Geddy Lee Rush Edmund Brown	Issue Number Issue Date	1 Feb. 6, 1995

11.0 Control of Inspection, Measuring, and Test Equipment

11.1 Scope and purpose

The quality system described in this section of the QAM conforms to the requirements of the standard(s): Element 4.11, Control of Inspection, Measuring, and Test Equipment.

11.2 Responsibility and authority (R&A)

The R&A for carrying out quality system activities have been assigned to the director of quality and the MR. All associates have the responsibility to carry out their work assignments in accordance with the quality policy and quality system documentation. The associates have been granted authority to complete the activities assigned in order to meet specified requirements.

11.3 Quality system requirements

Documented procedures controlling all aspects of the maintenance, control, and calibration of inspection, measuring, and test equipment (IM&TE) are in use. These procedures control the following activities:

- Hardware and software IM&TE are included.
- Methods and frequency of checking and rechecking acceptability and accuracy of IM&TE are utilized.
- Technical information supporting the acceptability and accuracy of IM&TE has been created and is available.
- Selection of IM&TE is based on analysis and determination of the precision required.
- IM&TE is identified by either a control equipment label or inscribed identifier number. The calibration status of each piece of equipment is determined by either a calibration sticker or record of calibration status, based on the equipment identification number.
- Calibration status of equipment is determined by either equipment vendor documentation or actual calibration of equipment by plant maintenance personnel.
- Calibration status for all IM&TE is traceable to industry, national, or international equipment standards.
- Methods and practices of calibration are documented, and these documents are adhered to in carrying out calibration activities.
- Documented procedures detail methods and practices to be used for assessing IM&TE found to be out of calibration and the actions related to product(s) that were inspected or tested using this equipment.

- Suitable environmental conditions are maintained to ensure the accurate operation of IM&TE and for product that will be inspected in these environmental conditions.
- Methods of handling, preservation, and storage exist to ensure that IM&TE is used in a manner that will ensure measurement accuracy and fitness for use.
- Measures have been taken to protect IM&TE from unauthorized adjustments that may affect the accuracy of the equipment.
- Records of IM&TE, including calibration records, are maintained.

11.4 Related and support documentation

2QP11-0001 IM&TE Hardware Procedure
2QP11-0002 IM&TE Software Procedure
2QP11-0003 IM&TE Identification & Calibration Status Practices
2QP11-0004 IM&TE Measurement System Analysis Practices and Methods

XYZ Typesetting Co. Quality Manual	Section 12.0 INSPECTION AND TEST STATUS	Page 1 of 1

12.0 Inspection and Test Status

12.1 Scope and purpose

The quality system described in this section of the QAM conforms to the requirements of the standard(s): Element 4.12, Inspection and Test Status.

12.2 Responsibility and authority (R&A)

The R&A for carrying out quality system activities have been assigned to the director of quality and the MR. All associates have the responsibility to carry out their work assignments in accordance with the quality policy and quality system documentation. The associates have been granted authority to complete the activities assigned in order to meet specified requirements.

12.3 Quality system requirements

Documented procedures controlling all aspects of inspection and test (I&T) status are in use. These procedures control the methods for identifying conforming and nonconforming products, based on their inspection and test results.

I&T status is maintained throughout the manufacturing process. No product is advanced until it has been determined that it is in conformance with specified requirements.

12.4 Related and support documentation

2QP12-0001 Inspection and Testing Status Procedure

Prepared By Approved By	George Mullin Oscar Stanage	Issue Number Issue Date	1 Feb. 6, 1995

XYZ Typesetting Co. Quality Manual	Section 13.0 CONTROL OF NON-CONFORMING PRODUCT	Page 1 of 1

13.0 Control of Nonconforming Product

13.1 Scope

The quality system described in this section of the QAM conforms to the requirements of the standard(s): Element 4.13, Control of Nonconforming Product.

13.2 Responsibility and authority (R&A)

The R&A for carrying out quality system activities have been assigned to the director of quality, director of manufacturing, and the MR. All associates have the responsibility to carry out their work assignments in accordance with the quality policy and quality system documentation. The associates have been granted authority to complete the activities assigned in order to meet specified requirements.

13.3 Quality system requirements

Documented procedures controlling all aspects of nonconforming product are in use. These controls include the following activities:

- Methods and practices used to identify and/or segregate
- Methods used to document characteristics and conditions
- Methods used to notify affected functional organizations
- Methods used to evaluate and carry out disposition, including the following actions:
  - Rework or repair
  - Reinspection
  - Regrading
  - Rejection
  - Scrapping
  - Acceptance through customer concession

13.4 Related and support documentation

2QP13-0001 Control of Nonconforming Product Procedure

2QP13-0002 Nonconforming Product Evaluation and Disposition Procedure

2QP13-0003 Temporary Deviation Practices

2QP13-0004 Rework and Repair Methods and Practices

2QP13-0005 Rejected Material—Scrap Procedures

Prepared By Approved By	Ted Striker Sally Decker	Issue Number Issue Date	1 Feb. 6, 1995

XYZ Typesetting Co. Quality Manual	Section 14.0 CORRECTIVE AND PREVENTIVE ACTION	Page 1 of 1

14.0 Corrective and Preventive Actions

14.1 Scope and purpose

The quality system described in this section of the QAM conforms to the requirements of the standard(s): Element 4.14, Corrective and Preventive Actions.

14.2 Responsibility and authority (R&A)

The R&A for carrying out quality system activities have been assigned to the director of quality, director of engineering, director of manufacturing, and the MR. All associates have the responsibility to carry out their work assignments in accordance with the quality policy and quality system documentation. The associates have been granted authority to complete the activities assigned in order to meet specified requirements.

14.3 Quality system requirements

Documented procedures controlling all aspects of corrective and preventive action (C&PA) are in use. These procedures may control the following:

- Determination of C&PA based on severity, magnitude, and risks
- Implementation of C&PA and modification of related documentation to reflect actions
- Use of methods and practices to address customer-related problems
- Use of methods to verify and report on customer satisfaction
- Determination of root cause(s) of nonconforming situations
- Identification and implementation of C&PA with intent of preventing recurrence of nonconforming situation
- Initiation and implementation of C&PA
- Maintenance of records associated with C&PA
- Submission of key information to management for review purposes

14.4 Related and support documentation

2QP14-0001 Corrective Action Procedure
2QP14-0002 Preventive Action Procedure
2QP14-0003 Customer Returned Goods Analysis Procedure

Prepared By Approved By	Henry David Ralph Waldo	Issue Number Issue Date	1 Feb. 6, 1995

XYZ Typesetting Co. Quality Manual	Section 15.0 HANDLING, STORAGE, PACKAGING, PRESERVATION, AND DELIVERY	Page 1 of 2

15.0 Handling, Storage, Packaging, Preservation, and Delivery

15.1 Scope and purpose

The quality system described in this section of the QAM conforms to the requirements of the standard(s): Element 4.15, Handling, Storage, Packaging, Preservation, and Delivery.

15.2 Responsibility and authority (R&A)

The R&A for carrying out quality system activities have been assigned to the director of material logistics and the MR. All associates have the responsibility to carry out their work assignments in accordance with the quality policy and quality system documentation. The associates have been granted authority to complete the activities assigned in order to meet specified requirements.

15.3 Quality system requirements

Documented procedures exist to control all aspects of handling, storage, packaging, preservation, and delivery (HSPP&D). These procedures are in use and control the following activities:

15.3.1 Handling. Handling methods and practices are intended to prevent damage and deterioration of material and products throughout the receiving, manufacturing, packaging, preservation, and shipping processes.

15.3.2 Storage. Receiving, in-process, and preshipment areas have been identified and are used. These areas have the intended purpose of preventing damage and deterioration to product(s) and/or material(s). Clearly defined methods and practices are in use for the receipt and dispatching of items from these areas.

At defined intervals, materials and products subject to degradation are evaluated and analyzed. Stock rotation (first in, first out) methods are in use. Acceptable inventory levels are maintained to meet specified requirements.

15.3.3 Packaging. Methods of packing, packaging, and marking of packaging materials are controlled to ensure that all specified requirements have been met.

Prepared By	Edgar Allen	Issue Number	1
Approved By	William Lee	Issue Date	Feb. 6, 1995

XYZ Typesetting Co. Quality Manual	Section 15.0 HANDLING, STORAGE, PACKAGING, PRESERVATION, AND DELIVERY	Page 2 of 2

15.3.4 Preservation. Measures are taken to preserve materials and products to prevent damage and deterioration.

15.3.5 Delivery. Practices and procedures are in use that provide for the protection of products after final inspection and testing. As required, this protection shall apply to the delivery of the product to the customer.

15.4 Related and support documentation

2QP15-0001 Handling Procedure
2QP15-0002 Storage Procedure
2QP15-0003 Packaging Procedure
2QP15-0004 Preservation Practices and Procedures
2QP15-0005 Delivery Procedures

Prepared By Approved By	Edgar Allen William Lee	Issue Number Issue Date	1 Feb. 6, 1995

XYZ Typesetting Co. Quality Manual	Section 16.0 CONTROL OF QUALITY RECORDS	Page 1 of 1

16.0 Control of Quality Records

16.1 Scope and purpose

The quality system described in this section of the QAM conforms to the requirements of the standard(s): Element 4.16, Control of Quality Records.

16.2 Responsibility and authority (R&A)

The R&A for carrying out quality system activities have been assigned to the director of quality and the MR. All associates have the responsibility to carry out their work assignments in accordance with the quality policy and quality system documentation. The associates have been granted authority to complete the activities assigned in order to meet specified requirements.

16.3 Quality system requirements

Documented procedures controlling all aspects of quality records are in use. These procedures control the following:

- Controls apply to all company and vendor-related quality records.
- Identification, collection, indexing, access, filing, storage, maintenance, disposal, disposition.
- Records are legible and maintained in a retrievable manner.
- Records are maintained in a storage environment that prevents damage and deterioration.
- Electronic records are backed up, via a tape backup system, and stored to prevent loss or damage.
- Record retention periods are specified and conform to customer requirements.

16.4 Related and support documentation

2QP16-0001 Control of Quality Records Procedure
2QP16-0002 Electronic Records Procedure
2QP16-0003 Retention Periods Procedure

Prepared By	Bill Wegman	Issue Number	1
Approved By	Bettina Ray	Issue Date	Feb. 6, 1995

17.0 Internal Quality Audits

17.1 Scope and purpose

The quality system described in this section of the QAM conforms to the requirements of the standard(s): Element 4.17, Internal Quality Audits.

17.2 Responsibility and authority (R&A)

The R&A for carrying out quality system activities have been assigned to the director of quality. All associates have the responsibility to carry out their work assignments in accordance with the quality policy and quality system documentation. The associates have been granted authority to complete the activities assigned in order to meet specified requirements.

17.3 Quality system requirements

Documented procedures controlling all aspects of internal quality audits (IQAs) are in use. These procedures control the following activities and requirements:

- IQAs are carried out to verify that planned and documented procedures, quality and control plans, and other quality system documentation are in conformity with the standard(s)
- IQAs are scheduled based on the department's impact on quality and quality performance.
- IQAs are carried out department by department, against all elements of the standard(s) that apply to the operation of the department being audited.
- IQAs are completed by trained and qualified personnel who understand the standard(s), auditing requirements, and basic communication skills.
- IQAs are carried out by personnel who are independent of the functional area being audited and free of bias or influence.
- IQA results are documented and are communicated to personnel responsible for the audited area.
- Auditee management will determine and implement timely corrective action.
- Follow-up activities are carried out to verify the effectiveness of IQA corrective action.
- Records of IQAs are maintained.

17.4 Related and support documentation

2QP17-0001 Internal Quality Audits Procedure
2QP17-0002 Work Environment Practices

XYZ Typesetting Co. Quality Manual	Section 18.0 TRAINING	Page 1 of 1

18.0 Training

18.1 Scope and purpose

The quality system described in this section of the QAM conforms to the requirements of the standard(s): Element 4.18, Training.

18.2 Responsibility and authority (R&A)

The R&A for carrying out quality system activities have been assigned to the director of human resources and the MR. All associates have the responsibility to carry out their work assignments in accordance with the quality policy and quality system documentation. The associates have been granted authority to complete the activities assigned in order to meet specified requirements.

18.3 Quality system requirements

Documented procedures controlling all aspects of training are in use. These procedures control the following:

- Training needs are identified based on individual job assignments and business needs.
- Training necessary to perform assigned jobs is provided.
- Individuals are qualified based on their abilities, on-the-job training, education, and other special skills.
- Grandfathering is permitted based on on-the-job training and experience. This information is documented.
- Records of training and qualifications are maintained.

18.4 Related and support documentation

2QP18-0001 New Employee Orientation Training Procedure
2QP18-0002 Training Needs Determination Procedure
2QP18-0003 Training Procedure

Prepared By Approved By	Tom Hardy Tess D'Urberville	Issue Number Issue Date	1 Feb. 6, 1995

XYZ Typesetting Co. Quality Manual	Section 19.0 SERVICING	Page 1 of 1

19.0 Servicing

19.1 Scope and purpose

The quality system described in this section of the QAM conforms to the requirements of the standard(s): Element 4.19, Servicing.

19.2 Responsibility and authority (R&A)

The R&A for carrying out quality system activities have been assigned to the director of quality and the MR. All associates have the responsibility to carry out their work assignments in accordance with the quality policy and quality system documentation. The associates have been granted authority to complete the activities assigned in order to meet specified requirements.

19.3 Quality system requirements

Documented procedures controlling all aspects of servicing are in use. These procedures control the following:

- Postsale servicing is performed when it is a specified requirement under a customer contract.
- Methods and practices are used to verify and report that the servicing meets the specified requirements.
- Records of postsale servicing activities are maintained.

19.4 Related and supporting documentation

2QP19-0001 Servicing Procedure

Prepared By	John Mayo	Issue Number	1
Approved By	Todd Nuavez	Issue Date	Feb. 6, 1995

XYZ Typesetting Co. Quality Manual	Section 20.0 STATISTICAL TECHNIQUES	Page 1 of 1

20.0 Statistical Techniques

20.1 Scope and purpose

The quality system described in this section of the QAM conforms to the requirements of the standard(s): Element 4.20, Statistical Techniques.

20.2 Responsibility and authority (R&A)

The R&A for carrying out quality system activities have been assigned to the director of quality and the MR. All associates have the responsibility to carry out their work assignments in accordance with the quality policy and quality system documentation. The associates have been granted authority to complete the activities assigned in order to meet specified requirements.

20.3 Quality system requirements

Documented procedures controlling all aspects of statistical techniques are in use. These procedures control the following activities and requirements:

- Determination of the need for statistical techniques
- Application of statistical techniques

20.4 Related and support documentation

2QP20-0001 Statistical Process Control Procedure
2QP20-0002 Statistical Process Control Practices

Prepared By	William Pilgrim	Issue Number	1
Approved By	Floyd Barber	Issue Date	Feb. 6, 1995

XYZ Typesetting Co. Quality Manual	Appendix A ORGANIZATION CHART	Page 1 of 1

Appendix A: Organizational Chart

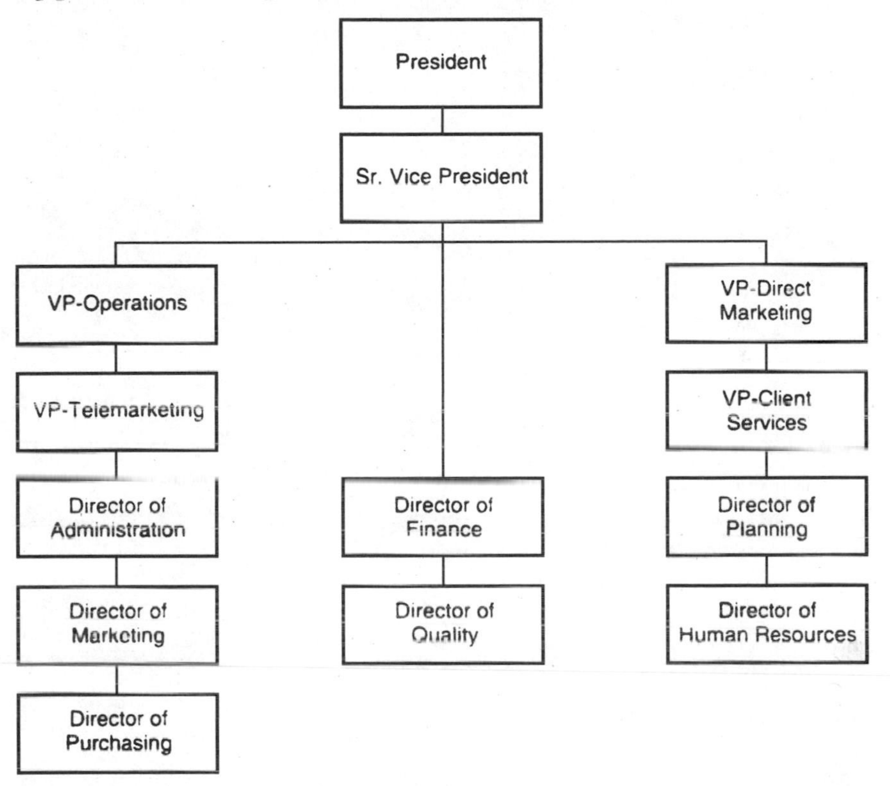

Prepared By	Walter Smilanovich	Issue Number	1
Approved By	Homer Winslow	Issue Date	Feb. 6, 1995

Bibliography

American Society for Quality. *ISO/CD1 9000: 2000, Quality Management Systems—Concepts and Vocabulary*. Milwaukee, WI, 1998.

American Society for Quality. *ISO/CD1 9001: 2000, Quality Management Systems—Requirements*. Milwaukee, WI, 1998.

American Society for Quality. *ISO/CD1 9004: 2000, Quality Management Systems—Guidelines*. Milwaukee, WI, 1998.

American Society for Quality. *ISO/DIS 9000: 2000, Quality Management Systems—Fundamentals and Vocabulary*. Milwaukee, WI, 1999.

American Society for Quality. *ISO/DIS 9001: 2000, Quality Management Systems—Requirements*. Milwaukee, WI, 1999.

American Society for Quality. *ISO/DIS 9004: 2000, Quality Management Systems—Guidelines for Performance Improvement*. Milwaukee, WI, 1999.

American Society for Quality. *ISO 9001: Quality Systems—Model for Quality Assurance in Design/Development, Production, Installation, and Servicing*. Milwaukee, WI, 1987.

American Society for Quality. *ISO 9003: Quality Systems—Model for Quality Assurance in Final Inspection and Test*. Milwaukee, WI, 1987.

American Society for Quality. *ISO 9002: Quality Systems—Model for Quality Assurance in Production and Installation*. Milwaukee, WI, 1987.

American Society for Quality. *ISO 9000: Quality Management and Quality Assurance Standards—Guidelines for Selection and Use*. Milwaukee, WI, 1987.

American Society for Quality. *ISO 9004: Quality Management and Quality System Elements—Guidelines*. Milwaukee, WI, 1987.

Automotive Industry Action Group. *Quality System Requirements—QS-9000,* 3d ed. Southfield, MI, 1998.

Automotive Industry Action Group. *Quality System Requirements—Tooling and Equipment Supplement,* 2d ed. Southfield, MI, 1998.

Barlas, Stephen. "U.S. Companies Feverishly Seek ISO 9000 Registration," *Managing Automation,* March 1992, p. 70.

Benson, Tracy E. "Quality Goes International," *Industry Week,* August 19, 1991, p. 55.

Berkman, Barbara N. "European Companies Join the Quality Crusade," *Electronic Business,* October 16, 1989, p. 263.

Costanzo, Anthony. "U.S. Corporate Executive Knowledge of ISO 9000 Lacking," *Quality,* September 1992, p. 47.

Graham, John F. "ISO-9000 Certification: Maintenance's Role," *Plant Services,* November 1991, p. 10.

Hutchens, Spencer. "Facing the ISO-9000 Challenge," *Compliance Engineering,* Fall 1991, p. 19.

International Organization for Standardization. *ISO 8402: Quality Management and Assurance—Vocabulary.* Geneva, Switzerland, 1994.
International Organization for Standardization. *ISO 9000: International Standards for Quality Management,* 2d ed. Geneva, Switzerland, 1992.
International Organization for Standardization. *ISO 9000-1: Quality Management and Quality Assurance Standards—Part 1: Guidelines for Selection and Use.* Geneva, Switzerland, 1994.
International Organization for Standardization. *ISO 9001: Quality Systems—Model for Quality Assurance in Design, Development, Production, Installation, and Servicing.* Geneva, Switzerland, 1994.
International Organization for Standardization. *ISO/CD2 9001: 2000, Quality Management Systems—Requirements.* Geneva, Switzerland, 1999.
International Organization for Standardization. *ISO 9002: Quality Systems—Model for Quality Assurance in Production, Installation, and Servicing.* Geneva, Switzerland, 1994.
International Organization for Standardization. *ISO/CD2 9004: 2000, Quality Management Systems—Guidance for Performance Improvement.* Geneva, Switzerland, 1999.
International Organization for Standardization. *ISO 9004-1: Quality Management and Quality System Elements—Part 1: Guidelines.* Geneva, Switzerland, 1994.
International Organization for Standardization. *ISO 9004-2: Quality Management and Quality System Elements—Part 2: Guidelines for Services.* Geneva, Switzerland, 1991.
International Organization for Standardization. *ISO 10011-1: Guidelines for Auditing Quality Systems—Part 1: Auditing.* Geneva, Switzerland, 1990.
International Organization for Standardization. *ISO 10011-2: Guidelines for Auditing Quality Systems—Part 2: Qualification Criteria for Quality Systems Auditors.* Geneva, Switzerland, 1991.
International Organization for Standardization. *ISO 10011-3: Guidelines for Auditing Quality Systems—Part 3: Management of Audit Programs.* Geneva, Switzerland, 1991.
International Organization for Standardization. *ISO/TS 16949: Quality Systems—Automotive Suppliers—Particular Requirements for the Application of ISO 9001: 1994.* Geneva, Switzerland, 1999.
International Organization for Standardization. *ISO/WD 1 19011: Guidelines on Quality and Environmental Auditing.* Geneva, Switzerland, 1999.
Johnson, Gary. "American Firms Face Challenge of Meeting New Quality Guidelines," *Denver Business,* June/July 1991, p. 44.
Johnson, Perry L. *Keeping Score: Strategies and Tactics for Winning the Quality War.* New York: HarperCollins, 1989.
Kiplinger Washington Letter, Vol. 69, No. 43, October 23, 1992.
Lamprecht, James L. *ISO 9000: Preparing for Registration.* New York: Marcel Dekker, 1992.
Niese, Ann. "ISO 9000: International Standard for Quality," *Electronic News,* November 1992.
Perry Johnson, Inc. *All About ISO 9000.* Text by Perry L. Johnson and Rob Kantner. Southfield, MI, 1992.
Perry Johnson, Inc. *How to Write Your ISO 9000 Quality Manual.* Text by Perry L. Johnson and Rob Kantner. Southfield, MI, 1992.
Placek, Chester. "Agreement on Standards, Testing, and Certification Outpacing Agreement on Many Other Issues Facing European Community," *Quality,* October 1991, p. 13.
Placek, Chester. "The ISO 9000 Edge," *Quality,* January 1992, p. 5.
QuEST Forum. *TL 9000 Quality System Requirements,* Book One, Release 2.5. New York, 1999.
Society of Automotive Engineers. *SAE Aerospace Basic Quality System Standard AS9000.* Warrendale, PA, 1997.

Society of Automotive Engineers. *SAE Aerospace Standard AS9100, Quality Systems—Aerospace—Model for Quality Assurance in Design, Development, Production, Installation and Servicing*. Warrendale, PA, 1999.

Stratton, John H. "What Is the Registrar Accreditation Board?" *Quality Progress*, January 1992, p. 67.

Tiratto, Joseph. "Preparing for EC 1992 in the US Through Quality System Registration," *Computer*, April 1991.

Verbrand der Automobilindustrie e.V. *VDA 6.1—Quality System Audit*, 4th ed. Frankfurt am Main, Germany, 1998.

Webb, Nanette M. "Flies in the Soup," *Quality in Manufacturing*, July/August 1991, p. 14.

Afterword

Here's one of the beauties of ISO 9000. When we went to Switzerland to assist in an audit over there, we sat down with QSF (a quality systems registrar and Swiss government quality body) to talk it over. And when we started talking about the ISO 9000 standard, we understood each other perfectly. Different countries, different cultures, different businesses, but with ISO 9000 there was no misunderstanding, no misinterpretation. We were talking the same language.

JIM ECKLEIN
Augustine Medical
Registered to ISO 9001,
EN 46001 and the
EU Medical Device Directive

Index

A

E

F

G

H

I

J

K

L

M

N

Q

R

S

T

About the Author

Perry L. Johnson is one of the world's top experts on ISO/QS-9000 and ISO 14000 consulting and training and a leading educator on the theories and practices of total quality management (TQM). A former Fortune 500 executive, he is globally recognized as an expert on ISO 9000, the international standard for quality, and on statistical and motivational methods to improve quality and productivity. In particular, Johnson is *also* known for his successful pioneering efforts to expand the use of statistical scorekeeping methodologies into short-run, administrative, and service operations.

A frequent speaker, panelist, and lecturer, Johnson is also the author of *Keeping Score: Strategies and Tactics for Winning the Quality War* (HarperCollins, 1989), *ISO 14000 Road Map to Registration* (McGraw-Hill, 1997), *ISO 14000: The Business Manager's Complete Guide to Environmental Management* (John Wiley & Sons, 1997), and *ISO/QS-9000 Yearbook: 1998* (McGraw-Hill, 1998). He has personally trained more than 1 million people in ISO 9000, QS9000, and TQM.